Inorganic Chemistry Concepts
Volume 2

Richard L. Carlin
A. J. van Duyneveldt

Magnetic Properties of Transition Metal Compounds

With 149 Figures
and 7 Tables

Springer-Verlag
New York Heidelberg Berlin

Richard L. Carlin
Department of Chemistry
University of Illinois at Chicago Circle
Chicago, Illinois 60680
USA

A. J. van Duyneveldt
Kamerlingh Onnes Laboratorium
University of Leiden
The Netherlands

Library of Congress Cataloging in Publication Data

Carlin, Richard Lewis, 1935–
Magnetic properties of transition metal compounds.

(Inorganic chemistry concepts ; v. 2)
Bibliography: p.
Includes index.
1. Transition metal compounds — Magnetic properties.
I. Duyneveldt, A.J. van, 1942– joint author. II. Title.
III. Series.
QD172.T6C32 546.6 77-18002

Printed in the United States of America.

9 8 7 6 5 4 3 2 1

ISBN 0-387-08584-X Springer-Verlag New York
ISBN 3-540-08584-X Springer-Verlag Berlin Heidelberg

To Dorothy ——

and Marijke

CONTENTS

PREFACE

This is a textbook of what is often called magnetochemistry. We take the point of view that magnetic phenomena are interesting because of what they tell us about chemical systems. Yet, we believe it is no longer tenable to write only about such subjects as distinguishing stereochemistry from the measurement of a magnetic susceptibility over a restricted temperature region; that is, paramagnetism is so well-understood that little remains to explore which is of fundamental interest.

The major purpose of this book is to direct chemists to some of the recent work of physicists, and in particular to a lengthy exposition of magnetic ordering phenomena. Chemists have long been interested in magnetic interactions in clusters, but many have shied away from long-range ordering phenomena. Now however more people are investigating magnetic behavior at temperatures in the liquid helium region, where ordering phenomena can scarcely be avoided.

The emphasis is on complexes of the iron-series ions, for this is where most of the recent work, both experimental and theoretical, has been done. The discussion therefore is limited to insulating crystals; the nature of magnetism in metals and such materials as semiconductors is sufficiently different that a discussion of these substances is beyond our purposes. The book is directed more at the practical experimentalist than at the theoretician.

Thus, this book tries to point the way that we believe the science of magnetochemistry should be going. In that regard, this book is very much an advertisement for the review article by L.J. de Jongh and A.R. Miedema which is referred to extensively throughout the book. RLC in particular has learned a great deal from this article and from conversations with Jos de Jongh, and would like to thank him for permission to depend so heavily on his thesis for our text.

We sincerely appreciate the secretarial assistance of Paula Hutton and Fran Petkus in the preparation of the manuscript. We would like to extend particular thanks to Yvonne Bosje, Astrid Durieux, Rita de Jong and Doky Lengkeek, who took care of typing the camera-ready copy, while at the same time carrying on their regular secretarial duties. We would also like to thank all the authors and copyright owners for permission to reproduce the published figures we have used. It was Wim Tegelaar who did all the work for the enormous number of figures throughout this book.

Chicago, Leiden R.L.C.
August 1977 A.J.v.D.

INTRODUCTION

The authors of this book asked me whether I was willing to make some historical remarks as an introduction. I gladly accepted this invitation as it is certainly useful and interesting to look back when realising how fast the development of physics in general and certainly also in the field of magnetic research was, and is. It is not my intention to give here a well-documented complete history but I shall restrict myself to some remarks from the period in which I worked in the Kamerlingh Onnes Laboratory of Leiden University. I entered this laboratory in 1929 after my "candidaats examen", more or less comparable with a B.Sc.

At the end of the 'twenties an important subject of research was still the measurement of susceptibilities of compounds down to about 14 K and the determination of the magnetic moments of the ions in the iron- and rare-earth groups. I remember De Haas telling me about the enthusiasm of Sommerfeld when De Haas told him at the Solvay Conference in 1930 about Gorter's measurements of the susceptibility of chromic alum which showed a constant "spin-only" moment till low temperatures. From these data the untenability of the theory of Laporte and himself about the moments of ions in the iron group became clear.

Some years before (in 1923) Woltjer had been able to give, by means of his measurements on Gd sulphate down to 1.3 K, a nice confirmation of Boltzmann's theory; these measurements were therefore, as Ehrenfest pointed out, of a basic importance. Because of the large value of the magnetic moment and the moderate accuracy of the measurements at that time the difference between a Brillouin function and a Langevin function was not evident. By means of the more accurate measurements of the paramagnetic rotation by Becquerel and this author, some 17 years later, it was shown that a satisfactory representation with a Langevin curve was not possible.

Other measurements by Woltjer, first with Kamerlingh Onnes, later with Wiersma, showed remarkable deviations from Curie's law for several anhydrous chlorides. They had some resemblance with ferromagnetism. This impression was strengthened by specific heat measurements of Schubnikow in Charkow, which also showed an anomaly. Just before the war these results were extended by measurements on a number of other compounds by Schultz and by Becquerel and the author, who spoke of metamagnetism. Also in other countries similar anomalous results were reported. Only later it became evident that these results were of an antiferromagnetic character. Though Néel had already put forward in 1932 a first indication of this new form of ordered magnetism, a suggestion which he developed further in 1936, only Van Vleck's article in 1941 made, after my impression, these ideas and their consequences more accessible, opening the doors to an immense field of research.

In 1939, a short time before the outbreak of the war, the first special magnetism conference took place in Strassbourg. I was already mobilized but obtained permission to replace J. Becquerel who was unable to come. The president of the congress was P. Weiss, professor in Strassbourg, which was a very important center of magnetic research. There were only some 30 official participants from 7 countries; about 9 others from these same countries took part in the discussions. Comparing this with the number of participants of the last magnetism conference, in 1976 in Amsterdam, that amounted to almost a thousand (980 from 34 countries), we see again the enormous increase of physical research. This Strassbourg conference can be considered as a closing of the first phase, and at the same time of the prewar period, a period in which the theoretical bases were drawn up and developed, and in which for instance the experimental application such as the obtaining of low temperatures by means of the adiabatic demagnetisation, as suggested by Debye and by Giauque and realised almost simultaneously in Berkeley and in Leiden, had also been further developed.

The postwar period was characterised by a rapid growth of the number of researchers and by the development of methods that had been discovered and worked out during the war for quite different purposes. I think here in particular of the application of cm waves used in EPR and also of other high-frequency techniques as necessary for NMR and magnetic relaxation. Also the rapid growth of the instrument industry, especially the electronics industry, may not be forgotten.

With respect to the Leiden research in particular one may, amongst many other subjects, think in the first place of the paramagnetic relaxation studies in which, under the supervision of C.J.Gorter, who started this research, very much work has been done. He had already done some work in this field before the war and during the war he wrote a fundamental monograph on it. Also the subject of antiferromagnetism can be mentioned. On $CuCl_2.2H_2O$, researches were carried out with a variety of methods, again under the leadership of C.J.Gorter. In a theoretical article, written in collaboration with the mathematician J.Haantjes, he presented a nicely rounded-off picture of a simple form of this phenomenon, based on the ideas of Néel.

The wealth of forms in which the magnetic phenomena present themselves to us nowadays always increases and therefore the study of them becomes more and more intricate. In order to prevent those who start doing research in the field of magnetism from soon getting off the track, good guides are necessary and I hope that the present book will be able to play a role as such.

J. van den Handel

CHAPTER I

PARAMAGNETISM: THE CURIE LAW

A. INTRODUCTION

This is a book concerning the magnetic properties of transition metal complexes. The subject has been of interest for a long time, for it was realized as long ago as the 1930's (1) that there was a diagnostic criterion between magnetic properties and the nature of the metal ion in a complex. Indeed, over the years, magnetic properties have continued to be used in this fashion. With time, the emphasis has changed, so that now chemists are becoming more interested in the magnetic phenomena themselves, and the subject is no longer a subsidiary one. One result of this new emphasis, which is hopefully rationalized and explained by this book, is that chemists must continue to decrease the working temperature of their experiments, with measurements at liquid helium temperatures now becoming common. In other words, the careful study of magnetic properties of transition metal complexes at low temperatures is essentially a redundant statement. Thus, little mention will be made of the many experimental results that pertain to high temperatures, that is, the temperatures of liquid nitrogen and above. The reason for this is simple, that the quantities which are being sought, such as the ground state energy levels, make a far more significant contribution to the measured phenomena at low temperatures. The emphasis is on measurements carried out on single crystals.

Magnetic susceptibilities and specific heats are among the principal measurements that chemists and physicists carry out on magnetic systems, and they therefore comprise the major part of the discussion. We hope to show how intimately connected the two are, and will attempt to correlate these magnetic properties with chemical structure.

Of course, one cannot, and should not, ignore the EPR spectra of metal ions in a work of this sort, but they are not discussed here at length. An introduction is presented, and some of the results are used, but the subject has already been discussed in many other books and reviews. Some subjects, such as neutron diffraction and nuclear resonance, are of special importance in magnetism, particularly in the study of ordered states, and some of the results obtained from these techniques are described here. Yet, these subjects cannot be treated in detail in a book of this length.

B. DIAMAGNETISM AND PARAMAGNETISM

The magnetic properties of transition metal compounds arise from the ground state of the metal ion as well as those thermally populated states. Thus, any energy level whose behavior in a field contributes to a magnetic susceptibility at a temperature T must lie within the order of kT of the ground state of the system. Conversely, the contribution of a particular level to the magnetic properties may in turn be altered by varying the temperature. The excited states of metal ions, which are of primary importance in determining the colors and optical spectra of these ions, are only of second order importance to magnetism.

Paramagnetic substances are those in which a magnetic field tends to align the magnetic moment of electron spins with the field. This will be discussed in detail below, but clearly the prerequisite for a sample to be paramagnetic is that the particular substance have unpaired electron spins. One result of this situation is that paramagnets are attracted into a magnetic field, and the fact that they therefore weigh more forms the basis of the familiar Gouy and Faraday balance experiments. The phenomenon of paramagnetism is a single-ion effect, by which we mean that we can explain paramagnetic behavior without invoking any interactions between the magnetic ions. Given the properties of an individual ion, we can calculate the properties of a mole of ions by a straightforward procedure, using Boltzmann thermal averaging. It is therefore appropriate to begin with some of the properties of the individual ions.

Briefly, atomic magnetism is due to the orbital motion (quantum number L) of the electrons, as well as the spin angular momentum (quantum number S). The resultant total angular momentum, J = L + S, is the important quantity, but L is often effectively zero, at least for the ground state of compounds of the iron series ions.

The majority of chemical substances is made up of ions or atoms with the noble gas configuration of filled electron shells, and it is found that L = S = J = 0, and only diamagnetism is exhibited. Diamagnetism is the property of being repelled by a magnetic field, and in the Gouy experiment, causes a sample to weigh less when a field is applied than in the absence of a field. There is a diamagnetic contribution even for paramagnetic substances, but it is usually small and can be accounted for by standard procedures.

Paramagnets are the subject of this book. The oxygen molecule, with two unpaired electrons, has S = 1, and is a well-known example. More relevant for our purposes are compounds containing such ions as Cr^{3+} and Mn^{2+} which have magnetic moments due to unpaired electron spins outside filled shells. The orbital motion is usually quenched by the ligand field, resulting in spin-only magnetism. Consider an isolated ion, acted on only by its diamagnetic ligands and an external magnetic field H (the Zeeman effect). The field will resolve the degeneracy of the various states according to the magnetic quantum number m, which varies from -J to J in steps of unity. Thus, the ground state of a free manganese(II) ion, which has L = 0, has S = 5/2 = J, and yields six states with m = ± 1/2, ± 3/2, and ± 5/2. These states are degenerate (of equal energy) in the absence of a field, but the magnetic field H resolves this degeneracy. The energy of each of the sublevels in a field becomes

$$E = mg\mu_B H, \tag{1.1}$$

where m is the magnetic quantum number and μ_B is the Bohr magneton,

$$\mu_B = \frac{|e|\hbar}{2mc} = 9.27 \times 10^{-21} \text{ erg/gauss}$$

and g is a (Landé) constant, characteristic of each system, which is equal to 2.0023 when L = 0, but frequently differs from this value. The convention used here in applying Eq (1.1) is that the ^6S ground state of the manganese ion at zero-field is taken as the zero of energy (2).

The situation is illustrated in Figure 1.1. The separation between adjacent levels, ΔE, varies with field, and is easily calculated as $\Delta E = g\mu_B H = 2\mu_B H$. In a small field of 1 kG,

$$\Delta E = 2\mu_B H = 2 \times 9.27 \times 10^{-21} \times 10^3 \approx 2 \times 10^{-17} \text{ erg}$$

while at a temperature of 1 K,

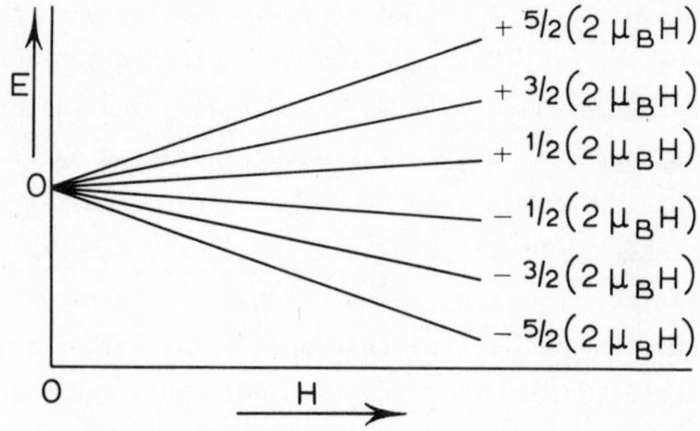

FIGURE 1.1 *Splitting of the lowest energy level of manganese(II) by a magnetic field into six separate energy levels.*

$$kT = 1.38 \times 10^{-16} \times 1 \approx 1.4 \times 10^{-16} \text{ erg.}$$

Thus, at H = 1 kG and T = 1 K, $\Delta E < kT$, and the resulting levels are populated almost equally, as may be found by calculating the distribution of magnetic ions among the various states from the Boltzmann relation,

$$N_i/N_j \propto \exp(-\Delta E_i/kT),$$

ΔE_i being the energy level separation between levels i and j.

Since each state corresponds to a different orientation with regard to the external magnetic field, the net magnetic polarization or magnetization M of the substance would then be very small. Or, the field tends to align the spins with itself, but this is opposed by thermal agitation.

On the other hand, in a large field of 20 kG, $\Delta E = 4 \times 10^{-16}$ erg, and at 1 K, $\Delta E > kT$. Then, only the state of lowest energy, $-Jg\mu_B H$, will be appreciably populated, having about 95% of the total. This corresponds to the ions lining up parallel to the external field, and the magnetization would almost have its largest or saturation value, M_{sat}. Real crystals that have been shown to exhibit behavior of this sort include chrome alum, $KCr(SO_4)_2 \cdot 12H_2O$, and manganese Tutton salt, $(NH_4)_2Mn(SO_4)_2 \cdot 6H_2O$. Each of the magnetic ions in these salts is well separated from the other magnetic ions, so it is said to be magnetically dilute. The behavior then is almost ideal-gas-like in the weakness of the interactions. The statistical mechanics of weakly interacting, distinguishable particles is therefore applicable. This means that most properties can be calculated by means of the Boltzmann distribution, as already suggested.

At low temperatures, the vibrational energy and heat capacity of everything but the magnetic ions may be largely ignored. The spins and lattice do interact through the time-dependent phenomenon of spin-lattice relaxation, a subject which will be introduced below. The magnetic ions form a subsystem with which there is associated a temperature which may or may not be the same as that of the rest of the crystal. The magnetization or total magnetic moment M is not correlated with the rest of the crystal, and even the external magnetic field has no effect on the rest of the crystal. Thus, the working hypothesis, which has been amply justified, is that the magnetic ions form a subsystem with its own identity, describable by the coordinates H, M, and T, independent of everything else in the system.

This model provides a basis for a simple derivation of the Curie law, which states that a magnetic susceptibility varies inversely with temperature. Although the same result will be obtained later by a more general procedure, it is useful to illustrate this calculation now. Consider a mole of spin-1/2 particles. In zero field, the two levels $m = \pm 1/2$ are degenerate, but split as illustrated in Figure 1.2 when a field $H = H_z$ is applied. The energy of each level is $mg\mu_B H_z$, which becomes $-g\mu_B H_z/2$ for the lower level, and $+g\mu_B H_z/2$ for the upper level; $\Delta E = g\mu_B H_z$, which for a g of 2, corresponds to about 1 cm^{-1} at 10000 gauss.

Now, the magnetic moment of an ion in the level n is given as $\mu_n = -\partial E_n/\partial H$; the molar macroscopic magnetic moment M is therefore obtained as the sum over magnetic moments weighted according to the Boltzmann factor.

$$M = N \sum_n \mu_n P_n = N \frac{\sum_{m=-1/2}^{1/2} (-mg\mu_B) \exp(-mg\mu_B H_z/kT)}{\sum_{m=-1/2}^{1/2} \exp(-mg\mu_B H_z/kT)}$$

where N is Avogadro's number and the summation in this case extends over only the two states, $m = -1/2$ and $+1/2$. Then,

$$M = \tfrac{1}{2} Ng\mu_B \left[\frac{\exp(g\mu_B H_z/2kT) - \exp(-g\mu_B H_z/2kT)}{\exp(g\mu_B H_z/2kT) + \exp(-g\mu_B H_z/2kT)}\right]$$

$$= \tfrac{1}{2} Ng\mu_B \tanh(g\mu_B H_z/2kT), \qquad (1.2)$$

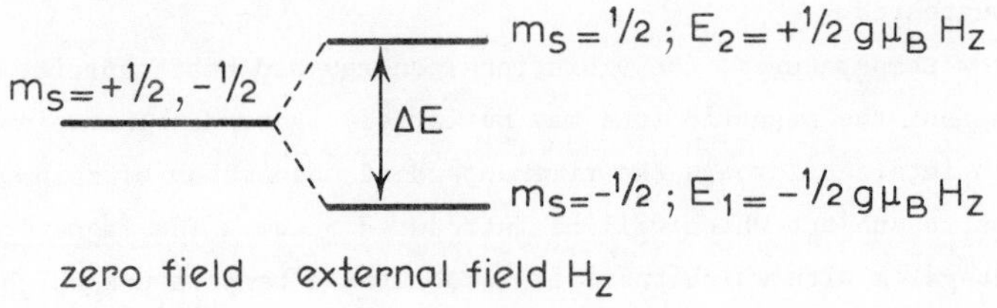

FIGURE 1.2 *Energy levels of an electron spin in an external magnetic field.*

since the hyperbolic tangent[†] is defined as

$$\tanh y = \frac{e^y - e^{-y}}{e^y + e^{-y}} \tag{1.3}$$

One of the properties of the hyperbolic tangent is that, for $y \ll 1$, $\tanh y = y$, as may be seen by expanding the exponentials:

$$\tanh y \approx \frac{(1 + y + \cdots)-(1 - y + \cdots)}{(1 + y + \cdots)+(1 - y + \cdots)} \approx y$$

and so for moderate fields and temperatures, with $g\mu_B H_z/2kT \ll 1$,

$$\tanh(g\mu_B H_z/2kT) \approx g\mu_B H_z/2kT$$

and thus

$$M = \frac{Ng^2\mu_B{}^2 H_z}{4kT} \tag{1.4}$$

Since the static magnetic susceptibility is defined as $\chi = M/H$, in this case

$$\chi \equiv M/H_z = \frac{Ng^2\mu_B{}^2}{4kT} = C/T, \tag{1.5}$$

[†]Some properties of the hyperbolic functions are described in the appendix.

which is in the form of the Curie law where the Curie constant

$$C = Ng^2\mu_B^2/4k.$$

This is a special case of the more general and more familiar spin-only formula,

$$\chi = \frac{Ng^2\mu_B^2 S(S + 1)}{3kT} \tag{1.6}$$

$$= \frac{N\mu_{eff}^2}{3kT}, \tag{1.7}$$

where $\mu_{eff}^2 = g^2\mu_B^2 S(S + 1)$ is the square of the "magnetic moment" traditionally reported by inorganic chemists. This quantity is of less fundamental significance than the static susceptibility itself, particularly in those cases where μ_{eff} is not independent of temperature. Other definitions of the susceptibility will be introduced in Chapt. II, since in practice one often measures the differential susceptibility, dM/dH, which is not always identical to the static one.

It is also of interest to examine the behavior of Eq. (1.2) in the other limit, of large fields and very low temperatures. In Eq. (1.3), if $y >> 1$, one may neglect e^{-y} compared to e^y, and

$$\tanh y = 1. \tag{1.8}$$

Then,

$$M = Ng\mu_B/2, \tag{1.9}$$

where the magnetization becomes independent of field and temperature, and, as discussed earlier, becomes the maximum or saturation magnetization which the spin system can exhibit. This situation corresponds to the complete alignment of magnetic dipoles by the field.

C. MAGNETIC MOMENT OF A MAGNETIC ION SUBSYSTEM

It is illuminating to calculate the magnetic moment of a magnetic system with arbitrary spin-quantum number . As above, with $J\hbar$ the total angular momentum, the energy is $E = -\vec{\mu}.\vec{H}$, where $\vec{\mu} = g\mu_B\vec{J}$; and

$E_m = mg\mu_B H_z$. The Boltzmann factor is

$$P_m = \exp(-mg\mu_B H_z/kT) \Big/ \sum_m \exp(-mg\mu_B H_z/kT)$$

so that $\langle\mu_z\rangle$, the average magnetic moment of one atom is

$$\langle\mu_z\rangle = \frac{\sum_{m=-J}^{J} (-mg\mu_B) \exp(-mg\mu_B H_z/kT)}{\sum_{m=-J}^{J} \exp(-mg\mu_B H_z/kT)}$$

But,

$$\sum_{m=-J}^{J} -mg\mu_B \exp(-mg\mu_B H_z/kT) = kT \frac{\partial Z_a}{\partial H_z}$$

where

$$Z_a = \sum_{m=-J}^{J} \exp(-mg\mu_B H_z/kT)$$

is the magnetic partition function for one atom. Hence,

$$\langle\mu_z\rangle = \frac{kT}{Z_a} \frac{\partial Z_a}{\partial H_z} = kT \frac{\partial \ln Z_a}{\partial H_z}$$

Now, define a dimensionless parameter η,

$$\eta = g\mu_B H_z/kT, \tag{1.10}$$

which measures the ratio of the magnetic energy $g\mu_B H_z$, which tends to align the magnetic moments, to the thermal energy kT, that tends to keep the system oriented randomly. Then,

$$Z_a = \sum_{m=-J}^{J} e^{-\eta m} = e^{\eta J} + e^{\eta(J-1)} + \cdots + e^{-\eta J}$$

which is a finite geometric series that can be summed to yield

$$Z_a = (e^{-\eta J} - e^{\eta(J+1)}) \;/\; (1 - e^{\eta})$$

On multiplying both top and bottom by $e^{-\eta/2}$, Z_a becomes

$$Z_a = \frac{e^{-\eta(J+\frac{1}{2})} - e^{\eta(J+\frac{1}{2})}}{e^{-\eta/2} - e^{\eta/2}} = \frac{\sinh\,(J+\frac{1}{2})\eta}{\sinh\,\eta/2},$$

since $\sinh y = \dfrac{e^{y} - e^{-y}}{2}$. Thus, $\ln Z_a = \ln\sinh\,(J+\frac{1}{2})\eta - \ln\sinh\,\eta/2$, and

$$<\mu_z> = \frac{kT\partial\ln Z_a}{\partial H_z} = \frac{kT\partial\ln Z_a}{\partial\eta}\;\frac{\partial\eta}{\partial H_z} = g\mu_B\,\frac{\partial\ln Z_a}{\partial\eta}$$

Hence

$$<\mu_z> = g\mu_B\left[\frac{(J+\frac{1}{2})\cosh(J+\frac{1}{2})\eta}{\sinh(J+\frac{1}{2})\eta} - \frac{\frac{1}{2}\cosh\,\eta/2}{\sinh\,\eta/2}\right]$$

or

$$<\mu_z> = g\mu_B J B_J(\eta) \tag{1.11}$$

where the Brillouin function $B_J(\eta)$ is defined as

$$B_J(\eta) = \frac{1}{J}[(J+\tfrac{1}{2})\,\coth(J+\tfrac{1}{2})\eta - \tfrac{1}{2}\,\coth\,\eta/2] \tag{1.12}$$

Now,

$$\coth y \equiv \frac{\cosh y}{\sinh y} = \frac{e^{y} + e^{-y}}{e^{y} - e^{-y}}$$

and for $y >> 1$, $e^{-y} << e^{y}$ and $\coth y = 1$. (A more exact result may also be obtained if desired; $\coth y = 1 + 2e^{-2y}$.) Conversly, for $y << 1$, both exponentials may be expanded and

$$\coth y = \frac{1 + \frac{1}{2}y^2 + \cdots}{y + \frac{1}{6}y^3 + \cdots} \approx (1 + \tfrac{1}{2}y^2)(1/y)\;/\;(1 + \tfrac{1}{6}y^2)$$

$$\approx (1/y)(1 + \tfrac{1}{2}y^2)(1 - \tfrac{1}{6}y^2) \approx (1/y)(1 + \tfrac{1}{3}y^2),$$

or for small y,

$$\coth y = 1/y + y/3. \tag{1.13}$$

Thus, in the two cases, the limiting behaviors of the Brillouin function are:

a) For $\eta >> 1$,

$$B_J(\eta) = 1/J[(J + \tfrac{1}{2}) - \tfrac{1}{2}] = 1$$

(Using the more exact expression for the coth, $B_J(\eta) = 1 - e^{-\eta}/J$)

$$\eta = g\mu_B H/kT >> 1 \text{ means } \frac{H}{T} >> \frac{k}{g\mu_B} = \frac{1.38 \times 10^{-16} \text{ erg/K}}{2 \times 9.27 \times 10^{-21} \text{ erg/G}} \approx 7 \text{ kG/K}$$

b) For $\eta << 1$,

$$B_J(\eta) = (1/J)\{(J + \tfrac{1}{2})[((J + \tfrac{1}{2})\eta)^{-1} + \tfrac{1}{3}(J + \tfrac{1}{2})\eta] - \tfrac{1}{2}[2/\eta + \eta/6]\}$$

$$= 1/J\,\{\tfrac{1}{3}(J + \tfrac{1}{2})^2\eta - 1/12\mu\} = (\eta/3J)(J^2 + J + \tfrac{1}{4} - \tfrac{1}{4})$$

$$= (J + 1)\eta/3 \tag{1.14}$$

and the initial slope of a plot of $B_J(\eta)$ vs. η will be $(J + 1)/3$. Such a plot for several values of J is illustrated in Figure 1.3.

For N non-interacting atoms, the mean magnetic moment, or magnetization, is

$$M = N \langle\mu_z\rangle = Ng\mu_B J B_J(\eta) \tag{1.15}$$

and so for η small, M is proportional to η, that is to say, $M \propto H/T$. In fact, for $g\mu_B H_z/kT << 1$, $M = \chi H_z$, where

$$\chi = Ng^2\mu_B^{\ 2} J(J + 1)/3kT \tag{1.16}$$

which once again is simply the Curie law. As one should now expect, in the limit of large η, $g\mu_B H_z/kT >> 1$, M becomes $Ng\mu_B J$, which is again the saturation moment. The physical ideas introduced here are not new, but of course the theory allows a comparison of theory with experiment over the

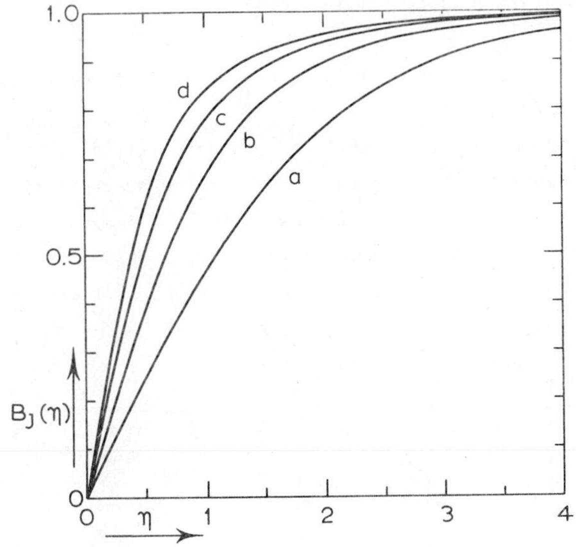

FIGURE 1.3 *The Brillouin function, $B_J(\eta)$, plotted vs. η for several values of J; curve a : J = 1/2, b : J = 3/2 c : J = 5/2 and d : J = 7/2.*

full range of magnetic fields and temperatures. The theory was tested by Henry (3) amongst others for three salts of different J value with no orbital contribution (i.e., g = 2.0) at four temperatures (1.30, 2.00, 3.00, 4.21 K). As illustrated in Figure 1.4, at 1.3 K a field of 50 kG produces greater than 99.5% magnetic saturation.

D. SOME CURIE LAW MAGNETS

The Curie constant

$$C = Ng^2\mu_B{}^2\, J(J + 1)/3k \tag{1.17}$$

takes the following form

$$C = \frac{[6.02 \times 10^{23}\ \text{mole}^{-1}][9.27 \times 10^{-21}\ \text{erg/G}]^2}{3\ (1.38 \times 10^{-16}\ \text{erg/K})}\ g^2 J(J + 1)$$

which, since it can be shown that gauss^2 equals erg/cm^3, becomes

$$C = 0.125\ g^2 J(J + 1)\ cm^3 K/mole$$

or dividing this expression by T in K, one has the volume susceptibility χ in units of $cm^3/mole$.

Since the question of the units of susceptibilities is often confusing, let us pursue the point further. For the susceptibility χ, the definition is $M = \chi H$, where M is the magnetization (magnetic moment per unit of volume) and H is the magnetic induction. This χ is dimensionless, but is expressed as emu/cm^3. The dimension of emu is therefore cm^3. The

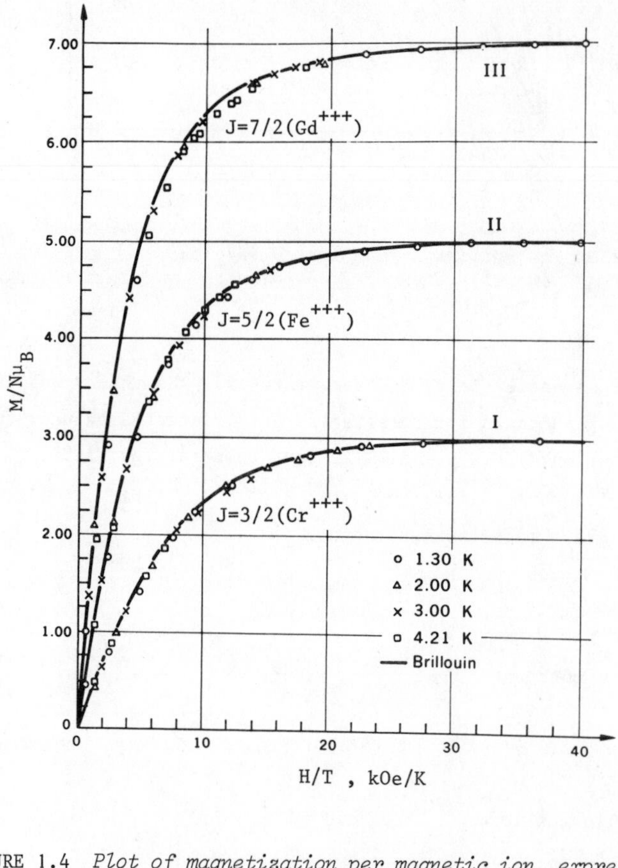

FIGURE 1.4 *Plot of magnetization per magnetic ion, expressed in Bohr magnetons, against H/T for (I) chromium potassium alum (J = 3/2); (II) iron ammonium alum (J = 5/2); and (III) gadolinium sulfate (J = 7/2). The points are experimental results of W.E. Henry (1952), and the solid curves are graphs of the Brillouin equation.*

molar susceptibility χ_N is obtained by multiplying χ with the molar volume, v (in cm^3/mole). So , the molar susceptibility leads to $M = H\chi_N/v$, or $Mv = \chi_N H$, where Mv is now the magnetic moment per mole. The dimension of molar susceptibility is thus emu/mole or cm^3/mole.

It should now be apparent that a good Curie law magnet will be found only when there are no thermally accessible states whose populations change with changing temperature. Three salts which offer good Curie-law behavior are listed in Table 1.1.

TABLE 1.1 Several Curie Law Magnets

	J	J(J+1)	C (expt) emu K/mole	C (calc) emu K/mole
$KCr(SO_4)_2 \cdot 12H_2O$	3/2	3.75	1.84	1.88
$NH_4Fe(SO_4)_2 \cdot 12H_2O$	5/2	8.75	4.39	4.38
$Gd_2(SO_4)_3 \cdot 8H_2O$	7/2	15.75	7.80	7.87

Each is assumed to have a g = 2.0, which is consistent with a lack of mixing of the ground state with nearby states with non-zero orbital angular momentum. The usual procedure for determining a Curie law behavior is

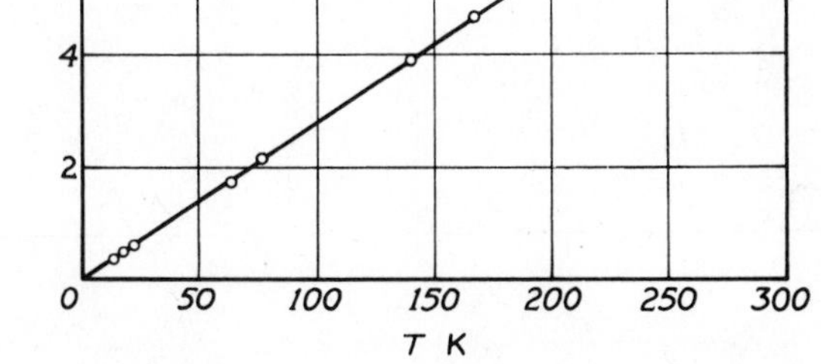

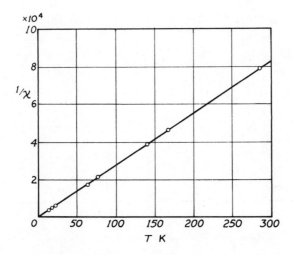

FIGURE 1.5 *The inverse susceptibility per gm of chromium potassium alum as a function of temperature, showing the very good agreement with Curie's law (after de Haas and Gorter, 1930).*

TABLE 1.2 The Paramagnetic Ions of the Rare Earth (4f) Group (4,5).

No. of electrons in 4f shell	Ion	State	S	L	J	g	Average C in units of $N\mu_B^2/3k$ exp	calc	Approx. energy (cm^{-1})
1	Ce^{3+}	$^2F_{5/2}$	1/2	3	5/2	6/7	6	6.43	
		$^2F_{7/2}$							2200
2	Pr^{3+}	$^3H_4$	1	5	4	4/5	12	12.8	
		$^3H_5$							2100
3	Nd^{3+}	$^4I_{9/2}$	3/2	6	9/2	8/11	12	13.1	
		$^4I_{11/2}$							1900
4	Pm^{3+}	$^5I_4$	2	6	4	3/5		7.2	
		$^5I_5$							1600
5	Sm^{3+}	$^6H_{5/2}$	5/2	5	5/2	2/7	2.4	0.71(2.5)	
		$^6H_{7/2}$							1000
6	Eu^{3+}	$^7F_0$	3	3	0	0	12.6	0(12)	
		$^7F_1$							400
7	Gd^{3+}	$^8S_{7/2}$	7/2	0	7/2	2	63	63	
		^6P							30000
8	Tb^{3+}	$^7F_6$	3	3	6	3/2	92	94.5	
		$^7F_5$							2000
9	Dy^{3+}	$^6H_{15/2}$	5/2	5	15/2	4/3	110	113	
		$^6H_{13/2}$							
10	Ho^{3+}	$^5I_8$	2	6	8	5/4	110	112	
		$^5I_7$							
11	Er^{3+}	$^4I_{15/2}$	3/2	6	15/2	6/5	90	92	
		$^4I_{13/2}$							6500
12	Tm^{3+}	$^3H_6$	1	5	6	7/6	52	57	
		$^3H_5$							
13	Yb^{3+}	$^2F_{7/2}$	1/2	3	7/2	8/7	19	20.6	
		$^2F_{5/2}$							10000

to plot χ^{-1} vs. T, resulting in a straight line with no intercept. Such a plot for chrome alum in Figure 1.5 illustrates how well the Curie law holds over a wide region of temperature for this substance.

E. SUSCEPTIBILITIES OF THE LANTHANIDES

Several new features may be introduced most easily by now considering some of the properties of the lanthanide ions. These ions have complete 5s and 5p shells and the magnetic properties are governed by the incomplete 4f shell. Because of size and screening effects, the 4f electrons do not interact strongly with the diamagnetic ligands, and the orbital moment is therefore not quenched. Thus, as the inorganic chemist is well aware, lanthanide ions in crystals do not differ greatly from the free ions, nor do many their properties vary from one compound to another nearly as markedly as do those of the iron series ions. Several properties are listed in Table 1.2. (The values in parentheses are those calculated by Van Vleck allowing for the population of excited states at T = 293 K).

The spectroscopic g factor used here is defined as

$$g = \frac{3J(J + 1) + S(S + 1) - L(L + 1)}{2J(J + 1)}$$

The relatively good agreement of the last two columns does not include Sm and Eu until, as Van Vleck showed, account is taken of the fact that there are low lying, thermally populated states.

Figure 1.6 illustrates the energy levels of an ion with L = 3 and S = 3/2. The different possible values of J (from L-S to L+S) are assumed to be separated by energies corresponding to several thousand Kelvins so that only the lowest level, J = 3/2, is effectively populated. In the free ion, (a), the fourfold degeneracy of this level is removed by a magnetic field, H. But, introduction of the ion into a crystalline lattice causes several effects, especially a crystal field splitting of some of the energy levels. A typical situation is illustrated in Figure 1.6b, where the lowest level may be split into two doublets. At low temperatures only the lowest doublet will be populated. The application of a magnetic field can then split the doublets as shown.

The Curie law is not found to be obeyed at low temperatures in some cases for lanthanide compounds. This is due to the splitting of the ground state levels in zero-field. An empirical $1/T^2$ correction to the

Curie law can be made in the fashion

$$\chi = \frac{C}{T}(1 - \Delta/T) \approx \frac{C}{T(1 + \Delta/T)} = \frac{C}{T + \Delta}$$

and a plot of $1/\chi$ vs. T should be linear with a non-zero intercept. This behavior is called a Curie-Weiss law.

Since the Curie law deviations are thought to be due to the resolution of ground state degeneracies by the crystalline field, it is important to have available a theorem that is due to Kramers. This is, that if the quantum number J is half-integral, then the crystalline field can at most only split the 2J + 1 states into half that number of doublets. Or,

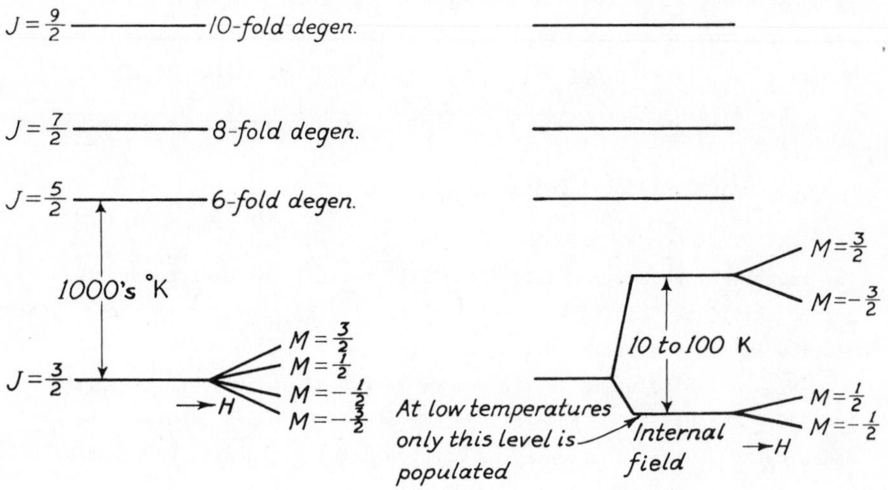

FIGURE 1.6 *The energy levels of an ion with L = 3 and S = 3/2. The different possible values of J running from L-S to L+S are separated by energies corresponding to several thousand Kelvins so that only the lowest level, J = 3/2, is populated. (a) In the free ion the fourfold degeneracy of this level is removed by a magnetic field, H. (b) In a salt the crystalline field can split this level into two doublets and at low temperatures only the lowest doublet will be populated. The application of a magnetic field splits the doublets as shown. From Ref. 8.*

to put it another way, systems with an odd number of electrons retain at least a two-fold degeneracy in the absence of a magnetic field. But, the significant point is that in systems with integral J (i.e., those with an even number of electrons) the states can be split into singlet levels of different energy and this may cause the ion to have a non-magnetic ground state. The levels need not be spaced evenly, and the order of the levels can also vary.

Typical situations are illustrated in Figure 1.7. There frequently is an overall energy spread of the levels of 10 to 100 K, or even more. These splittings can be smaller than kT at room temperature, and allow the Curie law to hold at these higher temperatures. The law breaks down at lower temperatures because the zero-field population of the levels changes.

Anisotropic crystals will often have anisotropic susceptibilities. The sources of magnetic anisotropy are varied and numerous, and much of this book concerns the information which can be gained from the measurements of anisotropic susceptibilities. It need hardly be stated that single crystals are necessary in order to examine this behavior. One source of anisotropy is simply the positioning of a (nominally) symmetric metal ion complex in a crystal of low symmetry; another source of anisotropy occurs with anisotropic complexes, such as a trans-MX_4Y_2 configuration. An axial crystalline field, in particular, tends to distort the electron cloud so that it too has axial symmetry. Spin-orbit coupling, in turn, causes the spins to be influenced by this orbital motion. An external field, however, also tends to align spins parallel to it, so that the spins tend to orient themselves along a resultant of the two fields of force they experience. The susceptibility is then anisotropic, and there will often be two main axes for the susceptibility - one parallel to the crystal field axis, and one perpendicular to it. The susceptibilities are usually denoted as $\chi_z = \chi_{\parallel}$ and $\chi_x = \chi_y = \chi_{\perp}$, although a rhombic component sometimes occurs, causing χ_y to differ from χ_x. An example is afforded by cerium ethylsulfate (Fig. 1.8), a hexagonal system, in which the susceptibilities are seen to individually follow a modified Curie law. The anisotropies vary from system to system, as do even the relative magnitudes. An important salt in low temperature physics is CMN, $Ce_2Mg_3(NO_3)_{12}$ $24H_2O$, for which $\chi_{\perp}/\chi_{\parallel} \approx 100$ at 1 K. The cerium ion, with S = 1/2 and L = 3, has a J = 5/2 ground state some 2200 cm^{-1} below the J = 7/2 state. The J = 5/2 level is split by the crystal field into three Kramers doublets. The splitting, Δ/k, between the lowest two doublets is approximately 36 K, so that effectively, at helium temperatures, only the lowest

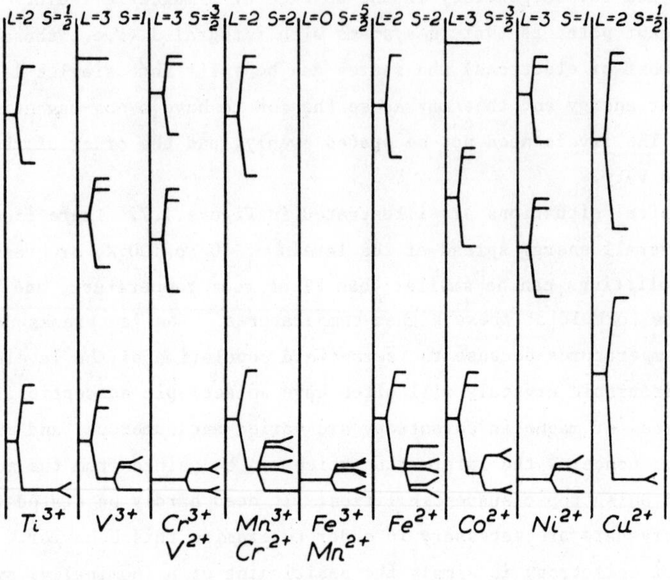
L=2 S=1/2
L=3 S=1
L=3 S=3/2
L=2 S=2
L=0 S=5/2
L=2 S=2
L=3 S=3/2
L=3 S=1
L=2 S=1/2
Ti3+
V3+
Cr3+
V2+
Mn3+
Cr2+
Fe3+
Mn2+
Fe2+
Co2+
Ni2+
Cu2+

(a)

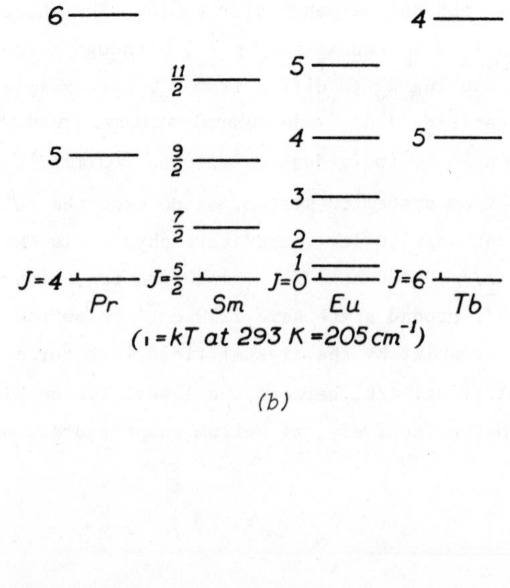
J=4
6
5
Pr
J=5/2
15/2
13/2
11/2
9/2
7/2
Sm
J=0
6
5
4
3
2
1
Eu
J=6
0
1
2
3
4
5
Tb
(1 = kT at 293 K = 205 cm-1)

(b)

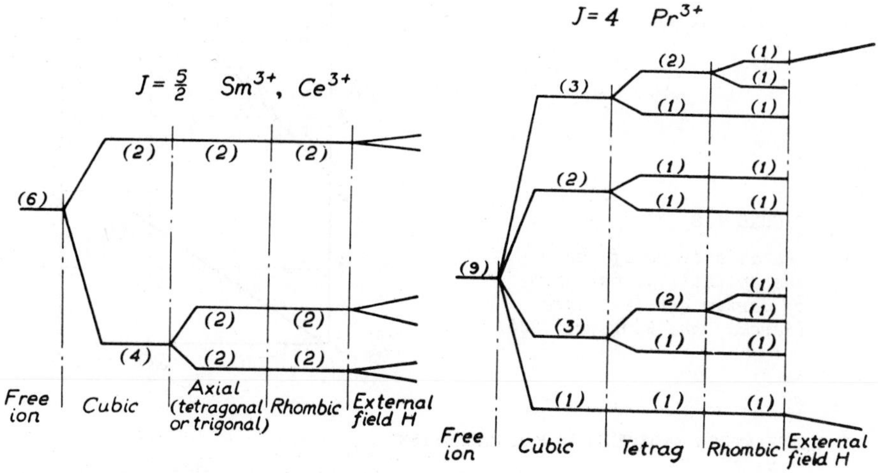

(c)

FIGURE 1.7 *Energy level diagrams. (a) The iron group showing how the 2L+1 levels are split, first by a cubic field, and then by a rhombic field. The manner in which the (2S+1)-fold degeneracy of these states is removed by spin-orbit coupling and by an external field is shown for the lowest level. These schemes are not to scale (Gorter, 1947). (b) The energy levels of four rare-earth ions. These are to scale and the energy corresponding to kT at 293 K is also shown. Each of these levels is (2J+1)-fold degenerate (Van Vleck, 1932). (c) The manner in which the lowest level of two of the ions in (b) is split by internal fields and by an external magnetic field, H. The amount of splitting and the sequence of levels, even in a cubic field, can change in different materials. For J = 4 the effect of H is only for the two extreme levels. The overall splitting might be of the order of 100 K. Not to scale. From Ref. 7.*

lying doublet is populated, and we are concerned with an effective spin-1/2 system (See Chapt. III-F for a fuller discussion of this point.) The splitting of the lowest doublet is described by an anisotropic g factor, with g perpendicular to the trigonal axis of 1.84, while $g_{\parallel} = 0.02$.

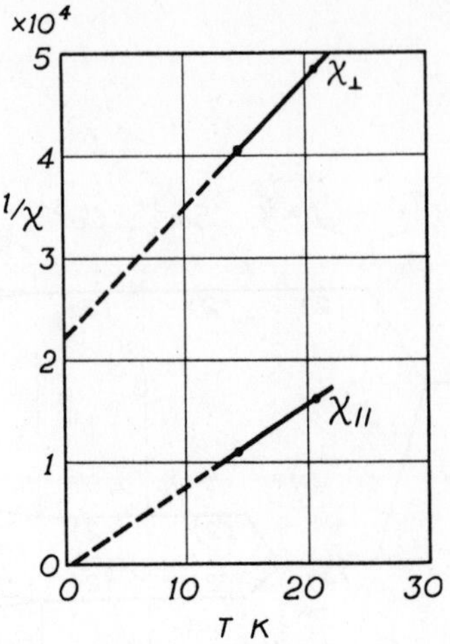

FIGURE 1.8
The anisotropy of the susceptibility, per gram, of cerium ethylsulphate (Fereday and Wiersma, 1935).

F. TEMPERATURE INDEPENDENT PARAMAGNETISM

It sometimes happens that systems with a spin singlet ground state, which from the development presented above would be expected to be diamagnetic, in fact exhibit a weak paramagnetic behavior. This paramagnetism is found to be temperature independent and, since it is only of the order of 10^{-4} cm^3/mole, is generally more important when considering measurements made at 80 K and above. This temperature independent paramagnetism (TIP) arises from a mixing into the ground state of the excited states that are not thermally populated. They may be connected with an orbital operator to the ground state, however; a spin operator will not suffice, for the problem assumes that S = 0 for the ground state, and all spin matrix elements will therefore be zero.

Consider a system whose ground state is both an orbital and spin singlet, such as the $^1A_1$ level of cobalt(III). Thus, calling the ground state $|0\rangle$,

$$\langle 0|\mu_z|0\rangle = 0$$

where $\mu_z = \frac{e}{2mc} L_z$ is the operator for the z component of orbital angular momentum. Let $|n\rangle$ be an excited state at an energy $\Delta = E_n - E_o$ above the ground state, and so the matrix element that connects the states is $\langle n|\mu_z|0\rangle$. Thus, in a weak field, μ_z mixes into the ground state some of the excited state and $|0\rangle$ becomes $|0'\rangle$, where

$$|0'\rangle = |0\rangle + \alpha|n\rangle$$

where α measures the amount of this mixing. More exactly, for $\mu_z H << \Delta$, by perturbation theory $|0'>$ becomes

$$|0'> = |0> + H \frac{<n|\mu_z|0>}{\Delta}|n>$$

To first order, the moment of the ground state becomes

$$<0'|\mu_z|0'> = \frac{2H|<n|\mu_z|0>|^2}{\Delta}$$

When $\Delta = E_n - E_o >> kT$, this has been called a high-frequency term, and the derived magnetization is

$$M = 2NH_z \frac{|<n|\mu_z|0>|^2}{\Delta}$$

or, summing over all states

$$\chi = \frac{M}{H} = 2N \sum_n \frac{|<n|\mu_z|0>|^2}{\Delta} = \frac{2}{3}N \sum_n \frac{|<n|\vec{\mu}|0>|^2}{E_n - E_o}$$

which will be found to be independent of temperature.

Let us apply the calculation to octahedral cobalt(III) (6,7), which has a $^1A_1$ ground state, a $^1T_1$ state at some 16000-21000 cm^{-1} above this, varying with the particular compound, and a $^1T_2$ state at some 24000-29000 cm^{-1}. There are two applicable symmetry restrictions. First, since μ_z is a one-electron operator, the excited states must involve a change in only one electron if they are to contribute. Cobalt(III) meets that restriction, for the $^1A_1$ state has a t_{2g}^6 configuration, while both the $^1T_1$ and $^1T_2$ states arise from the $t_{2g}^5e_g^1$ configuration.

Secondly, the orbital operator $\vec{\mu}$ (or $\vec{L}$) belongs to the T_{1g} symmetry of the octahedral group. Since the ground state is of A_{1g} symmetry, there must be available an excited state n of T_{1g} symmetry to allow a non-zero matrix element. (That is, the direct product $A_{1g} \times T_{1g} \times T_{1g}$ contains the A_{1g} irreducible representation.) This condition is also met here. Thus, one calculates (6,7)

$$<0|\sum_i L_{z_i}|n> = 2\hbar\sqrt{2}$$

so

$$\chi_{TIP} = \frac{2}{3} N \left(\frac{e\hbar}{2mc}\right)^2 \frac{24}{\Delta(T_1)} = 4.085/\Delta(T_1),$$

with $\Delta(T_1)$ in wave numbers. For $Co(NH_3)_6^{3+}$, the $^1T_{1g}$ state is at 21000 cm^{-1}, so one calculates

$$\chi_{TIP} = 1.95 \times 10^{-4}\ cm^3/mole,$$

which is the order of magnitude of the experimental result.

Temperature independent paramagnetism has been observed in such other systems as chromate and permanganate ions and with such paramagnetic systems as octahedral cobalt(II) complexes which have low-lying orbital states.

REFERENCES

1. L. Pauling, "Nature of the Chemical Bond," Cornell University Press, Ithaca, N.Y., 1940.
2. R.L. Carlin, J. Chem. Educ. 43, 521 (1966).
3. W.E. Henry, Phys. Rev. 88, 559 (1952).
4. B.I. Bleaney and B. Bleaney, "Electricity and Magnetism," Clarendon Press, Oxford, 1957.
5. A. Abragam and B. Bleaney, "Electron Paramagnetic Resonance of Transition Ions," Oxford University Press, Oxford, 1970.
6. C.J. Ballhausen and R. Asmussen, Acta Chem. Scand. 11, 479 (1957).
7. J.S. Griffith and L.E. Orgel, Trans. Faraday Soc. 53, 601 (1957).
8. H.M. Rosenberg, "Low Temperature Solid State Physics," Oxford U.P., Oxford, 1963.

GENERAL REFERENCES

H.B.G. Casimir, "Magnetism and Very Low Temperatures," Dover Publications, New York, 1961.

A. Earnshaw, "Introduction to Magnetochemistry," Academic Press, New York, 1968.

C.G.B. Garrett, "Magnetic Cooling," Harvard University Press, Cambridge, 1954.

E.S.R. Gopal, "Specific Heats at Low Temperatures," Plenum Press, New York, 1966.

J. van den Handel, "Handbuch der Physik, Bd XV," Springer Verlag, Berlin, 1956.

R.P. Hudson, "Principles and Application of Magnetic Cooling," North-Holland, Amsterdam, 1972.

C. Kittel, "Introduction to Solid State Physics," J. Wiley and Sons, New York, Ed. 4, 1971.

F.E. Mabbs and D.J. Machin, "Magnetism and Transition Metal Complexes," Chapman and Hall, London, 1973.

J.H. van Vleck, "The Theory of Electric and Magnetic Susceptibilities," Oxford U.P., Oxford, 1932.

M.W. Zemansky, "Heat and Thermodynamics," McGraw-Hill, New York, Ed. 5, 1968.

CHAPTER II

THERMODYNAMICS AND RELAXATION

A. INTRODUCTION

It is clear, from the preceding chapter, that the relative population of the energy levels of paramagnetic ions depends on both the temperature and the magnetic field. The equilibrium states of a system can often be described by three variables, of which only two are independent. For the common example of the ideal gas, such variables are PVT (pressure, volume, temperature). For magnetic systems one obtains HMT (magnetic field, magnetization, temperature) and the thermodynamic relations derived for a gas can be translated to a magnetic system by replacing P by H and V by -M. In the next section we review a number of the thermodynamic relations. Then, two sections are used to demonstrate the usefulness of these relations in analyzing experiments. In fact the simple thermodynamic relations are often applicable, even to magnetic systems that require a complicated model to describe the details of their behavior (1).

In the remaining part of this chapter it is then discussed what happens if an assembly of magnetic ions is not in equilibrium with its surroundings. The relaxation processes that then occur cause the differential susceptibility dM/dH, which is often the quantity of experimental interest, to vary considerably from its static value, M/H.

B. THERMODYNAMIC RELATIONS

The first law of thermodynamics states that the heat dQ added to a system is equal to the sum of the increase in internal energy dU and the work done by the system. For a magnetic system work has to be done on the system in order to change the magnetization, so the first law of thermodynamics may be written as

$$dQ = dU - HdM.$$

Remembering that the entropy S is related to Q by TdS = dQ, the first law can be written as

$$dU = TdS + HdM. \tag{2.1}$$

The energy U is an exact differential, so

$$\left(\frac{\partial T}{\partial M}\right)_S = \left(\frac{\partial H}{\partial S}\right)_M \tag{2.2}$$

The enthalpy E is defined as E = U - HM, thus

$$dE = dU - HdM - MdH = TdS - MdH,$$

and, as E is also an exact differential

$$\left(\frac{\partial T}{\partial H}\right)_S = -\left(\frac{\partial M}{\partial S}\right)_H \tag{2.3}$$

The Helmholtz free energy F is defined as F = U - TS, so

$$dF = dU - TdS - SdT = -SdT + HdM,$$

and the exact differentiability leads to

$$\left(\frac{\partial S}{\partial M}\right)_T = -\left(\frac{\partial H}{\partial T}\right)_M \tag{2.4}$$

The Gibbs free energy is G = E - TS, thus

$$dG = dE - TdS - SdT = -SdT - MdH$$

and

$$\left(\frac{\partial M}{\partial T}\right)_H = \left(\frac{\partial S}{\partial H}\right)_T \tag{2.5}$$

Equations (2.2) - (2.5) are the Maxwell relations in a form useful for magnetic systems.

The specific heat of a system is usually defined as dQ/dT, but it also depends on the particular variable that is kept constant when the temperature changes. For magnetic systems one has to consider both c_M and c_H, the specific heats at constant magnetization and field, respectively. From dQ = TdS and the definitions of U and E one obtains

$$c_M = \left(\frac{\partial Q}{\partial T}\right)_M = T\left(\frac{\partial S}{\partial T}\right)_M = \left(\frac{\partial U}{\partial T}\right)_M \tag{2.6}$$

and

$$c_H = \left(\frac{\partial Q}{\partial T}\right)_H = T\left(\frac{\partial S}{\partial T}\right)_H = \left(\frac{\partial E}{\partial T}\right)_H \tag{2.7}$$

Now let the entropy, which is a state function, be a function of temperature and magnetization, S = S(T,M). Then, an exact differential may be written

$$dS = \left(\frac{\partial S}{\partial T}\right)_M dT + \left(\frac{\partial S}{\partial M}\right)_T dM.$$

Multiplying through by T, and using the Maxwell relation given by Eq. (2.4),

$$TdS = T\left(\frac{\partial S}{\partial T}\right)_M dT - T\left(\frac{\partial H}{\partial T}\right)_M dM$$

in which the coefficient of dT is just the specific heat at constant magnetization. Thus

$$TdS = c_M dT - T\left(\frac{\partial H}{\partial T}\right)_M dM \tag{2.8}$$

In a similar way the entropy may be considered as a function of temperature and field, S = S(T,H). Then

$$dS = \left(\frac{\partial S}{\partial T}\right)_H dT + \left(\frac{\partial S}{\partial H}\right)_T dH,$$

and multiplying through by T and using Maxwell relation (2.5) yields

$$TdS = c_H dT + T\left(\frac{\partial M}{\partial T}\right)_H dH. \tag{2.9}$$

Specific heats are of interest in experimental work, and it is important to have expressions for the difference as well as for the ratio between the two specific heats, c_H and c_M, respectively. By subtracting Eq. (2.8) from Eq. (2.9) one obtains

$$(c_H - c_M)dT = -T\left(\frac{\partial M}{\partial T}\right)_H dH - T\left(\frac{\partial H}{\partial T}\right)_M dM,$$

and because

$$dT = \left(\frac{\partial T}{\partial H}\right)_M dH + \left(\frac{\partial T}{\partial M}\right)_H dM$$

a comparison of the coefficients of dM shows

$$\left(\frac{\partial T}{\partial M}\right)_H = \frac{-T\left(\frac{\partial H}{\partial T}\right)_M}{c_H - c_M}$$

From this equation one can resolve the quantity $c_H - c_M$. The fact that only two variables out of HMT are independently variable is expressed also by

$$\left(\frac{\partial H}{\partial T}\right)_M \left(\frac{\partial T}{\partial M}\right)_H \left(\frac{\partial M}{\partial H}\right)_T = -1, \tag{2.10}$$

a relation that may be used to eliminate $(\partial H/\partial T)_M$ from the expression for the difference between the two specific heats. So,

$$c_H - c_M = T\left(\frac{\partial M}{\partial T}\right)_H^2 \left(\frac{\partial H}{\partial M}\right)_T. \tag{2.11}$$

In a similar way an expression for the ratio between the two specific heats is obtained. Solving for dT from Eq. (2.9), then

$$dT = \frac{TdS}{c_H} - \frac{T}{c_H}\left(\frac{\partial M}{\partial T}\right)_H dH.$$

But also

$$dT = \left(\frac{\partial T}{\partial S}\right)_H dS + \left(\frac{\partial T}{\partial H}\right)_S dH$$

and the coefficients of dH must be equal. This leads to

$$c_H = -T\left(\frac{\partial M}{\partial T}\right)_H \left(\frac{\partial H}{\partial T}\right)_S$$

In an identical way, the use of Eq. (2.8) leads to

$$c_M = T\left(\frac{\partial H}{\partial T}\right)_M \left(\frac{\partial M}{\partial T}\right)_S$$

For the quotient c_H/c_M we derive

$$c_H/c_M = -\frac{\left(\frac{\partial M}{\partial T}\right)_H \left(\frac{\partial H}{\partial T}\right)_S}{\left(\frac{\partial M}{\partial T}\right)_S \left(\frac{\partial H}{\partial T}\right)_M} = -\frac{\left(\frac{\partial T}{\partial M}\right)_S \left(\frac{\partial H}{\partial T}\right)_S}{\left(\frac{\partial T}{\partial M}\right)_H \left(\frac{\partial H}{\partial T}\right)_M}$$

an expression that can be simplified by the use of relation (2.10)

$$c_H/c_M = (\partial M/\partial H)_T/(\partial M/\partial H)_S. \qquad (2.12)$$

The variation of M upon H is just the differential susceptibility. The constancy of either S or T determines whether it is the adiabatic or the isothermal susceptibility, respectively. These susceptibilities will be discussed in more detail in the last two sections of this chapter.

C. THERMAL EFFECTS

The specific heat of a magnetic system is, as we shall see repeatedly, one of its most characteristic and important properties. Magnetic ordering in particular, is evidenced by such thermal effects as the specific heat. Single ion anisotropies also offer characteristic heat capacity curves, as explained in Chapter III. We show here that there can even be a specific heat contribution to a Curie law paramagnet under certain conditions. But first, it is necessary to discuss lattice heat capacities.

Every substance, whether it contains ions with unpaired spins or not, exhibits a lattice heat capacity. This is because of the spectrum of lattice vibrations, which form the basis for both the Einstein and Debye theories of lattice heat capacities. For our purposes here, it is sufficient to be aware of the phenomenon and, in particular, that the lattice heat capacity decreases with decreasing temperature. It is this fact that causes so much of the interest in magnetic systems to concern itself with measurements at low temperatures, for then the magnetic contribution constitutes a much larger fraction of the whole.

In the Debye model the lattice vibrations (phonons) are assumed to occupy the 3N lowest energies of an harmonic oscillator. The Debye lattice specific heat, derived in any standard text of solid state physics, is

$$c_L = 9R(T/\theta_D)^3 \int_0^{\theta_D/T} \frac{e^x x^4}{(e^x-1)^2} dx$$

where $x = \hbar\omega/kT$ and $\theta_D = \hbar\omega_{max}/k$ is called the Debye characteristic temperature. At low temperatures, where x is large, the integral in the above expression becomes a constant; thus c_L may be approximated as

$$c_L \propto (T/\theta_D)^3$$

which is useful up to temperatures of the order of $\theta_D/10$. Each substance has its own value of θ_D, but as a practical matter many of the insulating salts which are the subject of this book obey a T^3 law up to approximately 20 K. The specific heat of aluminum alum, a diamagnetic salt which is otherwise much like many of the salts of interest here, has been measured (2) and does obey the law

$$c_L = 0.801 \times 10^{-3}\ T^3 \text{ cal/mole K}$$

at temperatures below 20K, as illustrated in Figure 2.1. On the other hand, substances which have clear structural features that are one or two dimensional in nature do not necessarily obey a T^3 law over wide temperature intervals, and caution must be used in assuming such a relationship.

Naturally, the lattice heat capacity must be evaluated in order to subtract it from the total to find the desired magnetic contribution. Several procedures are in common use:

1. Many of the magnetic contributions have, as we shall see below, a T^{-2} dependence for the specific heat in the high temperature limit. If the lattice specific heat follows the T^3 law in the measured temperature region, the total specific heat in a situation such as this should obey the relationship

$$c = aT^3 + bT^{-2} \qquad (2.13)$$

   and a plot of cT^2 vs. T^5 will, if it is linear, show the applicability of the procedure in the particular situation at hand over a certain temperature interval, as well as to allow the evaluation of the constants a and b. Extrapolation to lower temperatures then allows an empirical evaluation of the lattice contribution.
2. Many substances are not amenable to such a procedure, especially those which exhibit short-range order effects (Chapt. VI). This is because the magnetic contribution extends over a wide temperature interval, too

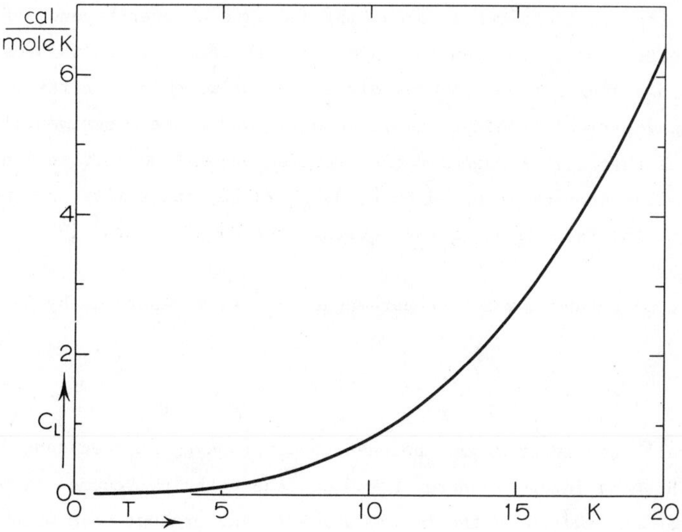

FIGURE 2.1 *Specific heat of* $KAl(SO_4)_2.12H_2O$. *Data from Ref. 2.*

wide and too high in temperature for Eq. (2.13) to be applicable. A procedure introduced by Stout and Catalano (3) is then often of value. The method depends on the law of corresponding states, which in the present situation states that the specific heats of similar substances will be similar, if weighted by the differences in molecular weights. In practice, one measures the total heat capacity of a magnetic system over a wide temperature region, and compares it to the heat capacity of an isomorphic but non-magnetic substance. By use of several relationships (3) one then calculates what the specific heat of the lattice of the magnetic compound is, and then subtracts this from the total. Though this procedure is used frequently it often fails just in those cases where it is needed the most. A careful analysis (4) suggests that the accuracy of this procedure is limited, especially in the application to layered systems, and a consistent evaluation of the lattice term can be made only in conjunction with an evaluation of the magnetic contribution.

3. Occasionally a diamagnetic isomorph will not be available. This is not a serious problem if the magnetic phenomena being investigated occur at sufficiently low temperatures but, again, if short range order effects are present, other procedures must be resorted to. Such a situation

occurs with $CsMnCl_3.2H_2O$, in which the broad peak in the magnetic heat capacity has a maximum at about 18K and extends even beyond 50K, where the lattice term is then the major contributor (5). In this case, the exchange constant (-3.3 K) was already known with some certainty, and so an empirical procedure could be used to fit the experimental data to the theoretical magnetic contribution and fitted lattice contribution. A similar problem is posed by $[(CH_3)_4N]MnCl_3$, and similar procedures led to the estimation of the several contributing terms (6).

Let us return now to the paramagnetic system described by Eq. (1.15)

$$M = N\langle\mu_z\rangle = Ng\mu_B J B_J(\eta)$$

and recall that interactions between the ions have not been considered so far. Thus the internal energy U = 0 and the enthalpy becomes simply the energy of the system in the field, which is the product of M with H_z. The heat capacity at constant field is obtained by differentiating the enthalpy with respect to temperature, Eq. (2.9), so that, for example, in the case of $J = \frac{1}{2}$, $B_{\frac{1}{2}}(\eta) = \tanh \eta/2$, $\eta = g\mu_B H_z/kT$

$$E = \frac{Ng\mu_B H_z}{2} \tanh(g\mu_B H_z/2kT)$$

and

$$c_H = \left(\frac{\partial E}{\partial T}\right)_H = \frac{Ng^2\mu_B^2 H_z^2}{4kT^2} \operatorname{sech}^2(g\mu_B H_z/2kT) \tag{2.14}$$

Eq. (2.14) is plotted in Figure 2.2 where it can be seen that a broad maximum occurs. The temperature range of the maximum can be shifted by varying the magnetic field strength. This curve, the shape of which is common for other magnetic phenomena as well, illustrates again the non-cooperative ordering of a paramagnetic system by the combined action of both magnetic field and temperature. In a number of cases $g\mu_B H_z/2kT \ll 1$ so the hyperbolic secant in Eq. (2.14) is equal to one, and the specific heat becomes

$$c_H = \frac{Ng^2\mu_B^2 H_z^2}{4kT^2} = \frac{CH^2}{T^2} \tag{2.15}$$

This simplification, using the Curie constant C, is in fact correct for more complicated systems also.

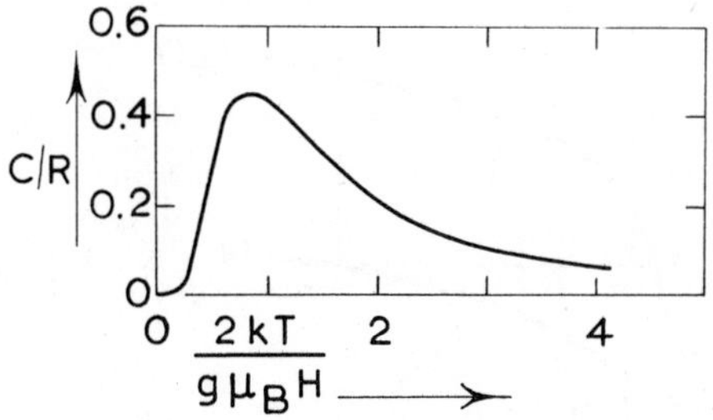

FIGURE 2.2 *Magnetic specific heat of a paramagnet, Eq. (2.14).*

D. ADIABATIC DEMAGNETIZATION

Until recently, adiabatic demagnetization served as the best procedure for obtaining temperatures below 1 K. The subject is introduced here because the exploration of appropriate salts for adiabatic demagnetization experiments was the initial impetus for much of the physicists' interest in paramagnets. Recall, in this regard, that kT changes by the same ratio whether the temperature interval be 0.1–1 K, 1–10 K, or even 10–100 K, and that the ratio of kT with some other quantity is often more significant than the particular value of T.

A schematic plot of the entropy of a magnetic system as a function of temperature for two values of the field, H = 0 and a nonzero H is illustrated in Figure 2.3. If for no other reason than the presence of the lattice heat capacity, every substance will have an entropy increasing with temperature in some fashion. A magnetic system will have a lower entropy, as a function of temperature, when the field is applied than when the field is zero, simply from the paramagnetic alignment caused by a field.

Consider that the system is already at temperatures of the order of 1 K, say at the point a on the S vs. T plot. Let the system be magnetized isothermally by increasing the field to the point b (Fig. 2.3). Application of the field aligns the spins, decreasing the entropy, and so heat is given off. The sample must be allowed to remain in contact with a heat sink at say, 1 K, so that the process is isothermal. The next step in the process requires an adiabatic demagnetization. The system is isolated from its surroundings and the magnetic field is removed adiabatically. The system moves horizontally from point b to point c, and the temperature is lowered significantly.

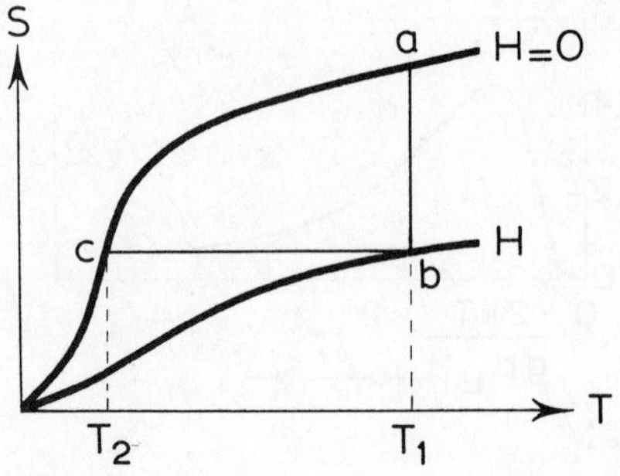

FIGURE 2.3 *Entropies (schematic) of a paramagnetic system with and without an external magnetic field.*

It is useful to examine these procedures in more detail with the help of the thermodynamic relations. Equation (2.9)

$$TdS = c_H\ dT + T\left(\frac{\partial M}{\partial T}\right)_H dH$$

simplifies for an isothermal process to

$$TdS = T\left(\frac{\partial M}{\partial T}\right)_H dH.$$

For a paramagnet, $M = \chi H = CH/T$, so $(\partial M/\partial T)_H$ will necessarily be negative, and thus dS will be negative for the first step. During the adiabatic demagnetization step ($dS = 0$), the above equation becomes

$$0 = c_H\ dT + T\left(\frac{\partial M}{\partial T}\right)_H dH$$

and since $(\partial M/\partial T)_H$ remains negative, for $dH < 0$, we find that $dT < 0$, and the system cools. (One should keep in mind that the heat capacity, c_H, is a positive quantity.) In general, a finite adiabatic change in field thus produces a temperature change given by

$$\Delta T = -\frac{T}{c_H}\left(\frac{\partial M}{\partial T}\right)_H \Delta H$$

if the adiabatic field change is 'ideal', that is, there is no heat exchange between the paramagnet and its surroundings. This effect is often called the magnetocaloric effect; it is the basis of adiabatic cooling, but also may cause unwanted temperature changes during experiments with pulsed magnetic fields. The method of adiabatic demagnetization stands on such a firm thermodynamic foundation that the first experiments by Giauque (7), one of the originators of the method, were not to test the method but in fact to use it for other experiments once the cooling occurred!

E. RELAXATION TIME AND TRANSITION PROBABILITY

In Chapter I the influence of a magnetic field on a system of magnetic ions was considered in so-called equilibrium situations only. In the preceding paragraph about adiabatic demagnetization it was demonstrated that the paramagnetic system behaves considerably different if it is isolated from its surroundings. Under experimental conditions isothermal or adiabatic changes often do not occur, so that the system will reach an equilibrium situation only after some time. In the following paragraphs we will concentrate on what happens in approaching the equilibrium situation.

The recovery of a perturbed magnetic system to a 'new' equilibrium can be described phenomenologically with the help of the relation:

$$\frac{dM}{dt} = (M_o - M)/\tau \tag{2.16}$$

where M_o is the equilibrium magnetization, M is the magnetization at time t, and τ is the relaxation time. In fact, the response times are very short, but, nevertheless, as soon as we study a magnetic material by means of techniques that require an oscillating field, the magnetization may no longer be able to follow the field changes instantaneously. One of the phenomena that has to be considered under such circumstances is the spin-lattice relaxation. This relaxation process describes the transfer of energy between the magnetic spin subsystem and the lattice vibrations.

Consider a mole of magnetic ions. In a magnetic field these magnetic ions may be assumed to have available two spin states, $|1\rangle$ and $|2\rangle$. Let the energy separation between these levels be Δ and the number of spins in each level n_1 and n_2, respectively. We are interested in spin-lattice relaxation, so the magnetic subsystem may be assumed to be in internal equilibrium at any time, and we define a spin temperature T_S by means of the Boltzmann relation,

$$n_2/n_1 = \exp(-\Delta/kT_S),$$

in which level $|1\rangle$ is assumed to be the lower in energy. The interaction between the lattice oscillations and the magnetic ions will cause transitions between the two energy levels. In our model

$$\frac{dn_1}{dt} = -\frac{dn_2}{dt} = -w_{12}n_1 + w_{21}n_2 \qquad (2.17)$$

where w_{12} represents the probability that a particle undergoes a transition from state $|1>$ to state $|2>$, etc. When the magnetic system is in thermal equilibrium with its surroundings(when temperature $T = T_S$), relation (2.17) is equal to zero, so

$$w_{21}/w_{12} = N_1/N_2 = \exp(\Delta/kT) \qquad (2.18)$$

(where the capital N_i's are introduced to indicate that the spins and lattice are in equilibrium).

We will show that the phenomenologically introduced time constant (Eq. (2.16)) is related to the above mentioned transition probabilities. Eq. (2.17) may be rewritten as

$$\begin{aligned} dn_1/dt = -dn_2/dt &= w_{21}(n_2-n_1) + w_{12}(n_2-n_1) + w_{21}n_1 - w_{12}n_2 \\ &= (w_{12}+w_{21})(n_2-n_1) + w_{21}(n_1-N_1) + w_{21}N_1 - w_{12}(n_2-N_2) - w_{12}N_2 \end{aligned}$$

The system consists of a mole of particles, thus

$$N_1 + N_2 = n_1 + n_2 = N$$

and

$$n_1 - N_1 = -n_2 + N_2,$$

so

$$\begin{aligned} dn_1/dt &= (w_{12}+w_{21})(n_2-n_1) - w_{21}(n_2-N_2) + w_{21}N_1 + w_{12}(n_1-N_1) - w_{12}N_2 \\ &= (w_{12}+w_{21})(n_2-n_1) - w_{21}n_2 + w_{21}N_2 + w_{21}N_1 + w_{12}n_1 \\ &\quad -w_{12}N_1 - w_{12}N_2. \end{aligned}$$

In this expression one may recognize, with opposite sign, the right-hand side of Eq. (2.17). We set these terms equal to dn_2/dt and subtract them from dn_1/dt, and try to obtain population differences in all terms. Then

$$d(n_1-n_2)/dt = (w_{12}+w_{21})(n_2-n_1) + w_{21}(N_1-N_2) + 2w_{21}N_2$$
$$+ w_{12}(N_1-N_2) - 2w_{12}N_1.$$

At thermal equilibrium $w_{21}N_2 = w_{12}N_1$, so that the above expression simplifies to

$$d(n_1-n_2)/dt = (w_{12} + w_{21})[(N_1-N_2) - (n_1-n_2)] \tag{2.19}$$

In a simple (symmetric) system with two energy levels the total magnetization M is directly proportional to the difference in occupation between the levels. In that case a comparison between the expressions (2.16) and (2.19) shows that

$$1/\tau = w_{12} + w_{21} \tag{2.20}$$

An expression similar to Eq. (2.20) can also be derived for a system with n energy levels (8).

F. SPIN-LATTICE RELAXATION PROCESSES

From the above it is clear that one has to calculate the transition probabilities w_{12}, etc., in order to predict the spin-lattice relaxation time constants. In performing such calculations one has to consider several possible processes:

1. Direct process. This is the relaxation process in which one magnetic ion flips to another energy level under the absorption or emission of the energy of one phonon. The frequency ω of the required phonon is determined by $\hbar\omega = \Delta$ if Δ is the energy change of the magnetic ion.
2. Raman process. In this non-resonant scattering process a phonon with frequency ω_1 is absorbed, causing the magnetic ion to reach a so-called virtual or non-stationary state from which it instantaneously decays by emission of a new phonon with frequency ω_2. The phonon frequencies are related to each other by $\hbar(\omega_1 - \omega_2) = \Delta$.
3. Orbach process. There is a possibility that a direct, resonant two-phonon process occurs via a real intermediate state, if the paramagnetic ion has, in addition to the two ground state levels, another level at such a position that the phonons can excite the ion to this state. In this case the phonon frequencies are also determined by $\hbar(\omega_1 - \omega_2) = \Delta$.

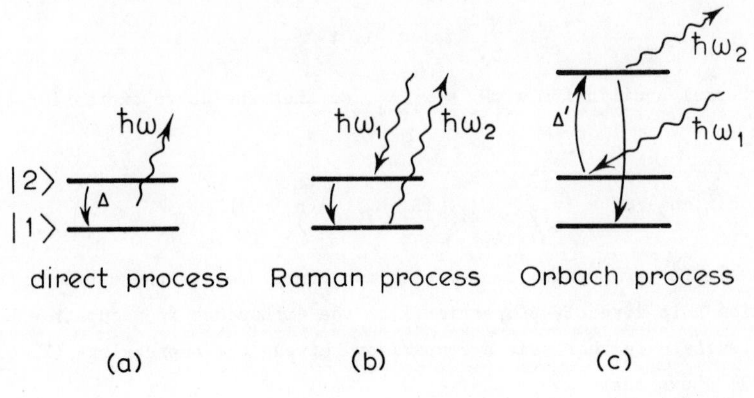

FIGURE 2.4 *The three spin-lattice relaxation processes indicated schematically.*

The three different types of relaxation processes are schematically indicated in Figure 2.4, where the curly arrows represent the phonons.

The spin-lattice interaction mechanism has to be considered in some detail if one wants to calculate the magnitudes of the time constants that characterize these processes as well as their dependence on external magnetic field and temperature. To give a complete derivation of all possible processes in various types of magnetic samples is far beyond the scope of this book (9). As example we give a slightly simplified derivation of the direct relaxation process.

Consider the transition of a spin from state $|2\rangle$ to state $|1\rangle$, accompanied by the creation of a phonon $\hbar\omega_p$:

Initial state of spin + phonon system $|I\rangle = |2\rangle|n_1 \ldots n_p \ldots\rangle$

Final state of spin + phonon system $|F\rangle = |1\rangle|n_1 \ldots n_{p+1} \ldots\rangle$,

where n_p represents the occupation number of the phonon mode ω_p.

If we write the state functions as above, we tacitly assume that the coupling between the spin and phonon systems is weak, thus allowing the factorisation as written. This assumption is valid as we will consider the model proposed by Van Vleck (10), in which the spin-phonon interaction is thought to occur via the electrical crystalline field, modulated by the phonons. In turn, the phonons modulate the orbital states which, by means of spin-orbit coupling, transfer the effect of the modulation to the spins. The important matrix element will be $\langle I|\mathcal{H}_{OL}^{(1)}|F\rangle$, and we have to consider the first order ion-phonon interaction $\mathcal{H}_{OL}^{(1)}$ for this case in more detail.

The crystal field potential at the site of a magnetic ion, V, can be expanded in terms of the relative displacements of the surrounding ions from their equilibrium positions, Δr_i:

$$V = V(\Delta r=0) + \sum_i \left(\frac{\partial V}{\partial \Delta r_i}\right)\Delta r_i + \tfrac{1}{2}\sum_{\substack{i,j\\ i\neq j}} \left(\frac{\partial^2 V}{\partial \Delta r_i \partial \Delta r_j}\right)\Delta r_i \Delta r_j + \ldots$$

The first order ion-phonon interaction may now be given as:

$$\mathcal{H}_{OL}^{(1)} = V - V(\Delta r=0) = \sum_i \left(\frac{\partial V}{\partial \Delta r_i}\right) \Delta r_i .$$

The relative displacements can be expressed as a function of creation and annihilation operators in the quantum mechanical description of the phonon spectrum (11). One finally arrives at:

$$\Delta r_i = \sum_p \sqrt{\omega_p}\, B_{ip}\, (a_p^+ - a_p)$$

in which a_p^+ and a_p are the creation and annihilation operators for phonons of energy $\hbar\omega_p$, respectively, and B_{ip} is a constant that contains a number of parameters, such as the crystal mass, the velocity of the phonons, etc. The explicit expression for B_{ip} can be found from Ref. 12. We will ignore the numerical constants and concentrate on the temperature and field dependence of the transition probabilities. Therefore, it is good to remember that the creation and annihilation operators fulfil the relations

$$\langle n_p | a_p | n_p + 1\rangle = (n_p + 1)^{\frac{1}{2}}$$

and

$$\langle n_p + 1 | a_p^+ | n_p\rangle = (n_p)^{\frac{1}{2}}$$

where n_p, the occupation number of the phonons in mode ω_p, is temperature dependent according to

$$n_p = [\exp(\hbar\omega_p/kT) - 1]^{-1}, \qquad (2.21)$$

as phonons obey Bose-Einstein statistics. The transition probability per unit time will be

$$w_{21} = (2\pi/\hbar)\,|\langle I|\mathcal{H}_{OL}^{(1)}|F\rangle|^2 \delta(E_F - E_I),$$

where the Dirac delta function is used to express the restriction of energy conservation.

We are interested in a spin that jumps from state |2> to |1>, creating a phonon $\omega = \Delta/h$. Considering the aspects of the perturbation hamiltonian, reviewed above, and omitting all constants that do not depend on ω or T, one obtains

$$w_{21} \propto \sum_f \sum_p \omega_p (n_p + 1) |\langle 2|V_f|1\rangle|^2 \delta(\omega_p - \Delta/\hbar) \qquad (2.22)$$

This expression is not of much use if we do not decide upon a model to describe the phonon system. In the simple Debye model the phonons are assumed to occupy the 3N lowest energies of an harmonic oscillator. In other words, the sum over the phonon modes becomes

$$\sum_p = \frac{V}{(2\pi)^3 v^3} \int_o^{\omega_D} \omega_p^{\ 2}\, d\omega_p \int d\Omega_k$$

where V is the volume of the crystal, v the velocity of sound in the crystal and Ω_k is a phase factor. We know that for instance neutron diffraction experiments have demonstrated departures from the Debye model in realistic systems (13), but for the description of relaxation phenomena these departures have minor effects. So in the calculation of transition probabilities one may use the Debye model and relation (2.22) becomes

$$w_{21} \propto \sum_f |\langle 2|V_f|1\rangle|^2 \int_o^{\omega_D} (n_p + 1)\, \omega_p^{\ 3}\, \delta(\omega_p - \Delta/\hbar) d\omega_p$$

which directly leads to

$$w_{21} = A \sum_f |\langle 2|V_f|1\rangle|^2 \Delta^3 \{[\exp(\Delta/kT) - 1]^{-1} + 1\}$$

if n_p is given by Eq. (2.22) and it is remembered that the phonon frequency is determined by $\hbar\omega = \Delta$. A tedious calculation gives the exact expression for the constant A. In a similar way one can derive that

$$w_{12} = A \sum_f |\langle 2|V_f|1\rangle|^2 \Delta^3 [\exp(\Delta/kT) - 1]^{-1}$$

so that

$$1/\tau = w_{12} + w_{21} = A \sum_f |\langle 2|V_f|1\rangle|^2 \coth(\Delta/2kT) \qquad (2.23)$$

For non-Kramers ions (even number of electrons) we can use for $|1\rangle$ and $|2\rangle$ the zero-th order state functions in the Zeeman perturbation which are temperature and field independent. This means that if it is allowed to consider $\Delta \propto H$, but still $\Delta < kT$, Eq. (2.23) simplifies to

$$1/\tau \propto H^2\ T \qquad (2.24)$$

The matrix elements in (2.23) are zero in first order if one considers the case of a Kramers ion (odd spin). Then admixtures of other states have to be taken into account, which causes the matrix elements to depend on H, and an extra H^2 dependence in the expression for the reciprocal relaxation time constant occurs, so that

$$1/\tau \propto H^4\ T \qquad (2.25)$$

In the following three equations we give the temperature and field dependences of the different kinds of relaxation processes. One has to realize that these relations are far from complete, and that the quoted dependences are simplified. The coefficients in each of the equations are different, and also vary from one magnetic substance to another.

$$1/\tau = A_1H^2T \quad + \quad B_1T^7 \quad + \quad C_1\ \exp(-\Delta/kT) \qquad (2.26)$$

$$1/\tau = A_2H^4T \quad + \quad B_2T^9 \quad + \quad C_2\ \exp(-\Delta/kT) \qquad (2.27)$$

$$1/\tau = A_3H^2T \quad + \quad B_3T^5 \qquad (2.28)$$

(Direct process) (Raman process) (Orbach process)

Eq. (2.26) refers to non-Kramers ions, Eq. (2.27) to Kramers ions with an 'isolated' doublet and Eq. (2.28) to Kramers ions with various doublets (energy difference between the doublets small compared to kT).

These equations show the characteristic features of spin-lattice relaxation processes. The terms representing the direct process depend on the magnetic field H, while the others do not. The reason for this different behavior is found in the assumed phonon distribution. In the direct process phonons of one particular frequency are involved, and the number of such phonons depends not only on temperature but also on the energy or,

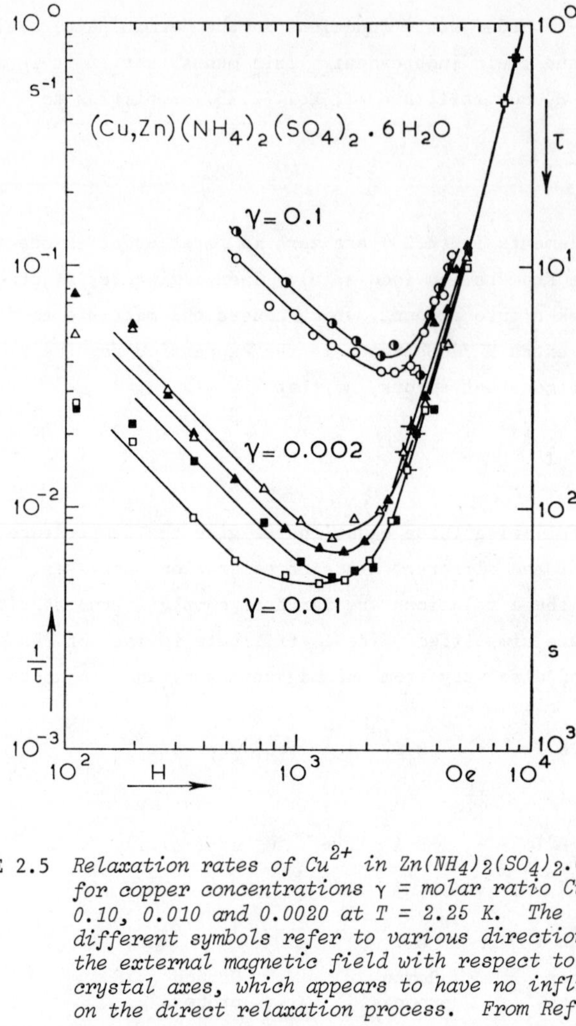

FIGURE 2.5 *Relaxation rates of* Cu^{2+} *in* $Zn(NH_4)_2(SO_4)_2 \cdot 6H_2O$ *for copper concentrations* γ = *molar ratio Cu:Zn* = *0.10, 0.010 and 0.0020 at* $T = 2.25$ *K. The different symbols refer to various directions of the external magnetic field with respect to the crystal axes, which appears to have no influence on the direct relaxation process. From Ref. 14.*

in other words, on the applied magnetic field. In the Raman and Orbach processes, phonons of all available energies participate as only the difference between two of them is important and the total number of phonons does not vary with the external magnetic field. A similar argument in which, apart from the features of the Debye model for the phonons, the characteristics of the Bose-Einstein statistics are also considered, gives a direct explanation for the more pronounced temperature dependences of the two-phonon relaxation processes. We now mention some experimental data in order to illustrate the characteristic field and temperature dependences

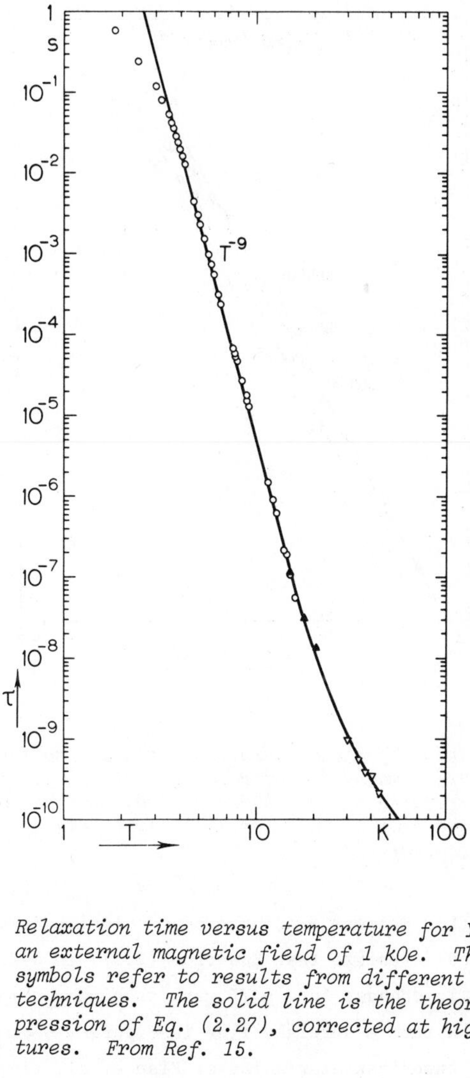

FIGURE 2.6 *Relaxation time versus temperature for $YbCl_3.6H_2O$ in an external magnetic field of 1 kOe. The various symbols refer to results from different experimental techniques. The solid line is the theoretical expression of Eq. (2.27), corrected at high temperatures. From Ref. 15.*

of τ as given in Equations (2.27) and (2.28).

The paramagnetic spin-lattice relaxation behavior of Cu^{2+} ions doped into $Zn(NH_4)_2(SO_4)_2.6H_2O$ is determined by a Kramers doublet, about 10^4 K below the other states. For samples with small Cu concentration, de Vroomen et al. (14) demonstrated that the relaxation rate τ^{-1} follows the H^4 dependence of the direct process at strong fields (Fig. 2.5). The temperature dependence of the Raman process is usually observed at fields of a few hundred oersteds. A typical example is given in Figure 2.6 where some

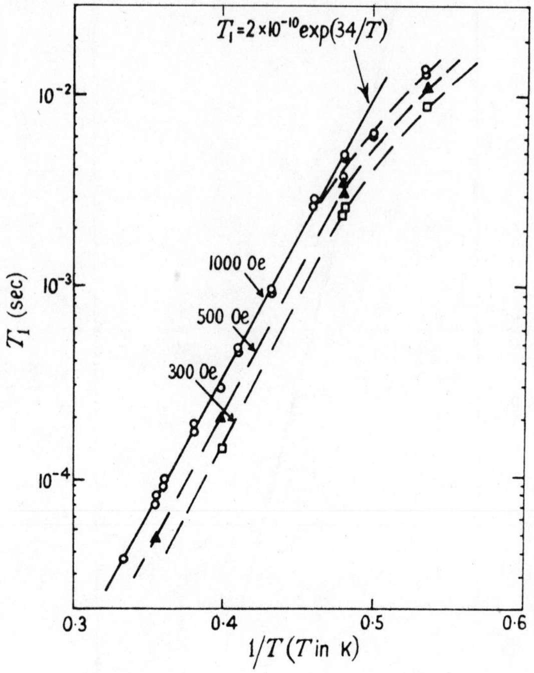

FIGURE 2.7 *A plot of the measured relaxation time for Ce^{3+} in cerium magnesium nitrate against 1/T for field strengths of 300, 500 and 1000 Oe. The solid line is a plot of $T_1 = 2 \times 10^{-10} \exp(34/T)$. From Ref. 16.*

of the data on another Kramers ion, ytterbium chloride hexahydrate, are collected. One should realize that Equation (2.27) is a simplification; the T^9 term gradually changes into a T^2 term at higher temperatures, which causes the deviation from a straight line in Figure 2.6 at 40 K.

The beautiful experimental results of Finn et al. (16) that led to the introduction of the so-called Orbach process are given in Figure 2.7. Cerium ions in cerium magnesium nitrate have a doublet state lowest with the next state at 36 K above (see also Chapter VIII-B-11). Note that in Figure 2.7 the relaxation time τ (or T_1) is plotted against 1/T (not T!), so that the exponential temperature dependence appears as a straight line with a slope of Δ/k.

As a final example in this section we concentrate on manganese ions (Chapt. VIII-B-5) which usually follow Eq. (2.28) as the doublets

± 1/2, ± 3/2 and ± 5/2 are only of the order of 0.1 K apart. Figure 2.8 shows the Raman relaxation process in Cs_3MnCl_5, varying as $\tau \sim T^{-5}$ around $T \simeq 10$ K (17). The relaxation processes of $Mn(NH_4)_2(SO_4)_2 \cdot 6H_2O$ were studied in the Leiden group of Prof. Gorter for over 30 years. Data are presented in Figure 2.9, which show the direct process in this compound as measured in the early seventies (18). The H^{-2} dependence of τ is present at liquid helium as well as at liquid hydrogen temperatures, but only if the external

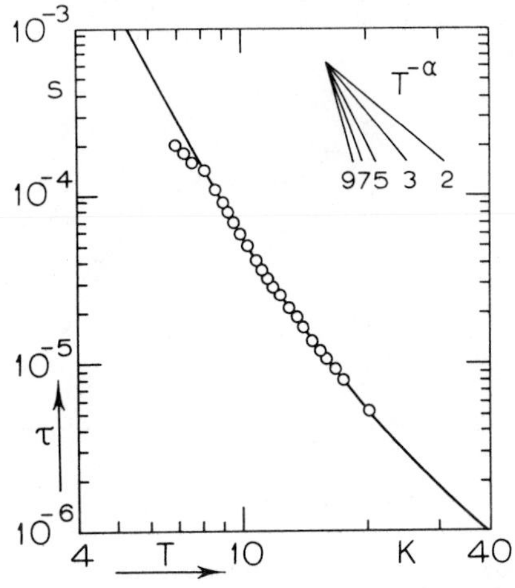

FIGURE 2.8 *The Raman relaxation time as a function of temperature for Cs_3MnCl_5 in an external field of 2 kOe. The solid line shows the theoretical temperature dependence of τ with $\tau \propto T^{-5}$ around 10 K and the curvature towards $\tau \propto T^{-2}$ at higher T. From Ref. 17.*

magnetic field exceeds 10 kOe. A longitudinal field of 10 kOe is accessible today for almost every experimentalist, but in fact the lack of this 'strong' field value, needed to fulfil the condition of isolated ions, is the reason why it took the experimentalists almost 30 years to verify the field dependence of the direct process relaxation time in manganese compounds. Several review articles on the theory of spin-lattice relaxation phenomena have been published recently (9,13). An interesting number of experiments is mentioned in the book of Standley and Vaughan (19), while Vaughan also published an article covering the literature published between 1971 and mid-1974 (20), so that recent experimental information can be obtained from these references.

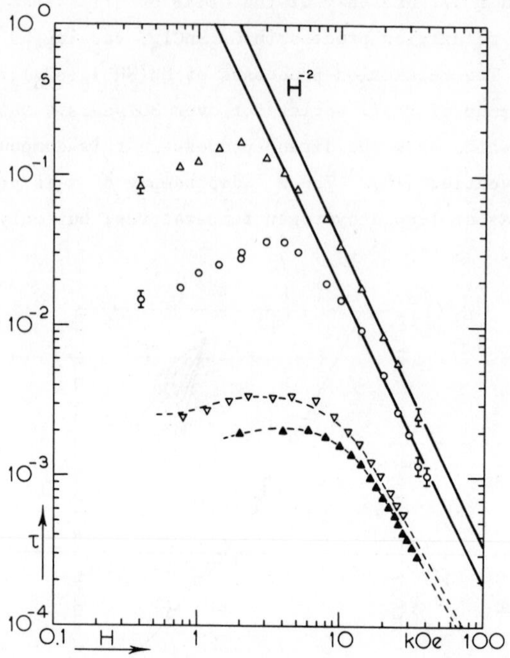

FIGURE 2.9 *Relaxation time versus external magnetic field for* $Mn(NH_4)_2(SO_4)_2 \cdot 6H_2O$. *Δ:T = 2.09 K; o:T = 4.22 K; ∇:T = 14.2 K; ▲:T = 16.0 K. The solid lines show the direct process at liquid helium temperatures with the characteristic* H^{-2} *dependence as measured above 10 kOe. The deviations at lower fields show the influence of relaxation processes other than those of Eq. (2.28). At the higher temperatures the two processes of Eq. (2.28) are operating in parallel, as can be seen from the computer fits through the data (dashed curves). From Ref. 18.*

G. SUSCEPTIBILITY IN ALTERNATING FIELDS

Now that we have introduced the phenomenon of spin-lattice relaxation it is interesting to consider its effect on the differential susceptibility. The static susceptibility that was introduced in deriving the Curie law, in Chapter I, was defined as M/H. In a number of experiments one actually measures the dynamic or differential susceptibility, dM/dH, by means of applying an oscillating magnetic field. From the preceding paragraph it will be clear that a system of magnetic ions is not always capable of following the changes of an external magnetic field immediately. In other words the redistribution of the magnetic spins over the energy levels proceeds via a relaxation process characterized by a time constant, τ, as given by one of the Eqs. (2.26) - (2.28). In practice it will make a difference whether one period of an oscillating field takes a long or a short time compared to the time constant τ, as in the latter case the magnetic spins are unable to redistribute in an optimal or equilibrium way over their energy levels.

Let us consider the frequency dependence of the differential susceptibility in more detail by returning to the system of N magnetic ions with two possible spin states $|1>$ and $|2>$ as used in Section E, above. Equation (2.18) defined the ratio between w_{21} and w_{12}. Let us write these transition probabilities as

$$w_{12} = w \exp(-\Delta/2kT)$$

and

$$w_{21} = w \exp(\Delta/2kT).$$

We also introduce $\delta n = n_1 - n_2$, and then $N + \delta n = 2n_1$, $N - \delta n = 2n_2$ and $dn_1/dt = -dn_2/dt = \frac{1}{2}\, d\delta n/dt$. With these new variables, Eq. (2.17) can be written as

$$\tfrac{1}{2}\, d\delta n/dt = -\tfrac{1}{2}(N + \delta n)\, w \exp(-\Delta/2kT) + \tfrac{1}{2}(N - \delta n)\, w \exp(\Delta/2kT)$$

We assume relatively high temperatures in order to simplify the mathematics, so $\Delta/2kT << 1$. The exponentials may then be approximated as $1 \pm \Delta/2kT$, thus

$$\tfrac{1}{2} d\delta n/dt = Nw\Delta/2kT - w\delta n \tag{2.29}$$

Solving Eq. (2.29) for Δ and taking the time derivative, one obtains

$$d\Delta/dt = (2kT/wN)(w\, d\delta n/dt + \tfrac{1}{2}\, d^2\delta n/dt^2) \tag{2.30}$$

As mentioned above, measurements of the differential susceptibility are performed by adding a small oscillating field, h, to a constant external one, H_o (with h and H_o parallel). In this case the applied magnetic field may be given by $H(t) = H_o + h \cos \omega t$, which by means of Euler's theorem

($\exp(i\omega t) = \cos \omega t + i \sin \omega t$), can also be written as

$$H(t) = H_o + \text{Re}\left[h \exp(i\omega t)\right]$$

in the complex notation. The magnetic ions will try to reach an equilibrium distribution over the energy levels in accordance with the value of the magnetic field. It seems plausible to assume the difference in the population numbers of the energy levels will oscillate also

$$\delta n = \delta n_o + \text{Re}\left[y \exp(i\omega t)\right]$$

The effect of spin-lattice relaxation will be apparent in the parameter y. The magnetic ions will be in equilibrium at the time scale that is used, if the oscillations in H are slow compared to τ, so y is maximal. On the other hand a large value for the angular frequency ω will imply a small value for y. Since we have to obtain the relation between y and the differential susceptibility let us first find an expression for y. From the above it follows that

$$d\delta n/dt = i\, y\, \omega \exp(i\omega t),$$

and

$$d^2\delta n/dt^2 = -y\, \omega^2 \exp(i\omega t).$$

With these expressions Eq. (2.30) reads as

$$d\Delta/dt = (2kT/wN)[\, iw\, y\, \omega \exp\, (i\omega t) - \tfrac{1}{2}y\omega^2 \exp(i\omega t)]\, .$$

Thus

$$y = (d\Delta/dt)(wN/2kT)\, [\, (i\omega w - \tfrac{1}{2}\omega^2) \exp(i\omega t)]^{-1} \tag{2.31}$$

The magnetization of an ion in state $|1\rangle$ will be equal to

$$m_1 = -dE_1/dH = +dE_2/dH = -m_2,$$

the second half of the relation being valid because the system is assumed to be symmetric, $E_1 = -E_2$. The total magnetization M = 0 if the energy levels are equally populated. In other cases

$$M = \delta n\, m_1 = \delta n\, dE_2/dH = \tfrac{1}{2}\delta n\, d\Delta/dH,$$

which in fact means that M changes due to variations in δn and in $d\Delta/dH$. (Remember that $\Delta = E_2 - E_1$ which is now $2E_2$.)

One may also assume

$$M = M_o + \mathrm{Re}\,[m \exp(i\omega t)],$$

if the system is linear, and that will be the case in most experiments. Thus

$$\begin{aligned} dM/dt &= i\omega m \exp(i\omega t) \\ &= \tfrac{1}{2}(d\Delta/dH)\, d\delta n/dt + \tfrac{1}{2}\delta n_o (d^2\Delta/dH^2)\, dH/dt \\ &= \tfrac{1}{2}(d\Delta/dH)\, i\omega y \exp(i\omega t) + \tfrac{1}{2}\delta n_o (d^2\Delta/dH^2)\, hi\omega \exp(i\omega t); \end{aligned}$$

in the second term of this expression one does not consider the time derivative of δn, so it is possible to assume $\delta n \simeq \delta n_o$. The differential susceptibility will now be

$$\chi = dM/dH = m/h = \tfrac{1}{2}(d\Delta/dH)y/h + \tfrac{1}{2}\delta n_o (d^2\Delta/dH^2),$$

an expression that shows the relation between y and χ. The physical impact of this equation can be seen better if we use Eq. (2.31) to eliminate y, and the time derivative of H to remove h from the expression. The result may then be given as

$$\chi = (N/4kT)(d\Delta/dH)^2 \left(1 + \frac{i\omega}{2w}\right)^{-1} + \text{terms of the order } (d^2\Delta/dH^2) \qquad (2.32)$$

The first term in this expression is frequency dependent. Together with ω the relaxation time τ appears in the expression, as from Eq. (2.20), it becomes equal to $1/2w$ because we assumed relatively high temperatures. The influence of the frequency of oscillation, ω, on the measured differential susceptibility is therefore directly related to the relaxation time. As stated above, the magnetic ions are to be considered in equilibrium as the fluctuations in H occur slowly compared to the time constant, τ. More specifically, this supposition means that $\omega\tau \ll 1$, so the quotient

$1/(1 + i\omega\tau) \simeq 1$ and χ becomes

$$\chi \simeq (N/4kT)\ (d\Delta/dH)^2 + \text{order}(d^2\Delta/dH^2) \qquad (2.33)$$

This value of the susceptibility has to be identical to the static susceptibility if our picture is correct. If we assume $\Delta = g\mu_B H$, then $(d\Delta/dH) = g\mu_B$ and $(d^2\Delta/dH^2) = 0$, so the low frequency limit of the differential susceptibility in Eq. (2.33) becomes

$$\chi = \frac{Ng^2\mu_B^{\ 2}}{4kT}$$

and that is indeed the value derived in Chapter I for the static susceptibility of an $S = \frac{1}{2}$ paramagnet, Eq.(1.5). In the following sections of this chapter we will call this low frequency limit of the differential susceptibility the isothermal susceptibility, χ_T, thus expressing the fact that the spins maintain thermal contact with the surroundings in an effective way. (One should realize that all of the susceptibilities mentioned in Chapter I are isothermal ones.)

Another interesting limit is obtained if one measures a differential susceptibility with an oscillating field of high frequency, e.g., in a situation that the magnetic ions are far from being capable of redistributing in accordance with H. So consider the case that $\omega\tau$ in Eq. (2.32) becomes much larger than 1. Then the first term of this equation diminishes and

$$\chi = \text{order}(d^2\Delta/dH^2) \qquad (2.34)$$

This situation represents an assembly of magnetic spins uncoupled from its surroundings. In other words, relation (2.34) represents the so-called adiabatic susceptibility, χ_{ad}; the value of χ_{ad} is strongly field dependent, and in fact at strong fields, $\chi_{ad} \to 0$. We will come back to this behavior in the next section.

One can introduce the two susceptibilities defined above in Eq. (2.32),

$$\chi = (\chi_T - \chi_{ad})/(1 + i\omega\tau) + \chi_{ad} \qquad (2.35)$$

Relation (2.32) can also be obtained for more complicated systems than our example, but the necessary algebra increases considerably. Another approach is followed in the original work of Casimir and Du Pré (21), who derived

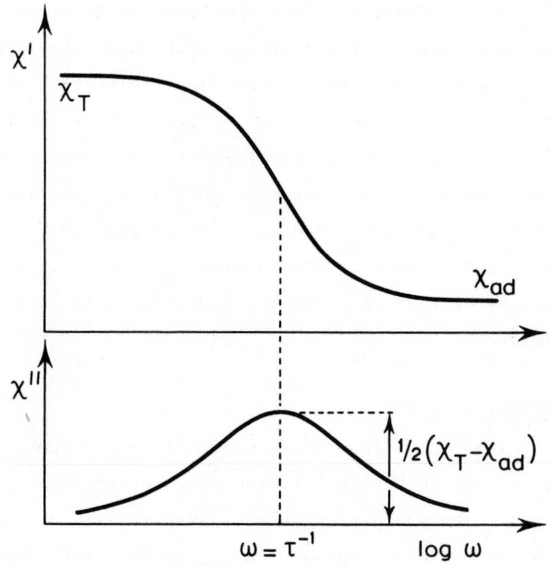

FIGURE 2.10 *Frequency dependence of* χ' *and* χ'' *according to Eq. (2.36).*

Eq. (2.35) on thermodynamic grounds. For the properties discussed in this book it is enough to consider the general expression (35) which gives an adequate description of the frequency dependence of the differential susceptibility as long as it is possible to characterize the spin-lattice relaxation process with a single time constant, τ.

Coming back to the experiments, one should realize that in Eq. (2.35) the differential susceptibility is described as a complex quantity. If one chooses $\chi = \chi' - i\chi''$, then

$$\chi' = (\chi_T - \chi_{ad})/(1 + \omega^2\tau^2) + \chi_{ad} \tag{2.36a}$$

and

$$\chi'' = \omega\tau(\chi_T - \chi_{ad})/(1 + \omega^2\tau^2) \tag{2.36b}$$

The frequency dependence of χ' and χ'' is schematically drawn in Figure 2.10. The two susceptibility limits that were introduced above display the in-phase component χ', $\chi' = \chi_T$ if $\omega \ll \tau^{-1}$ and $\chi' = \chi_{ad}$ if $\omega \gg \tau^{-1}$. The out-of-phase component χ'' approaches zero at these limits, but shows a maximum around the frequency $\omega = \tau^{-1}$, the height of the maximum being $\frac{1}{2}(\chi_T - \chi_{ad})$. This maximum provides a method for determining τ, which is the basis for non-resonance spin-lattice relaxation experiments. Figure 2.10 demonstrates the difficulties in comparing a differential susceptibility with the simple models of Chapter I. It is not enough to know whether such a susceptibility possesses an out-of-phase component or not, as one still may be measuring χ_{ad} instead of χ_T. In the next section we will consider the differences between χ_{ad} and χ_T in more detail. Some experimental data on relaxation time constants are reviewed in Sect. F. so one may find there some of the situations in which it is necessary to be concerned about relaxation behavior.

H. ADIABATIC SUSCEPTIBILITIES

The ratio between the adiabatic susceptibility and the isothermal one was shown in Eq. (2.12) to be equal to the ratio between the specific heats, c_M and c_H. In measuring the differential susceptibility at high frequencies the in-phase component χ' approaches χ_{ad} as was seen from Eq. (2.36a). It is necessary to evaluate the actual ratio χ_{ad}/χ_T in more detail, because as was mentioned above, χ_{ad} is strongly field dependent and goes to zero at strong fields.

The specific heat at constant magnetization, c_M, was shown in Section B to be equal to $(\partial U/\partial T)_M$ in which U represents the internal energy. One has to consider the influences of zero-field splittings and of interactions (dipolar- and exchange-) between the magnetic ions in order to find U. These effects will be described in some detail in the following chapters. For our present purpose it is enough to know that in a number of cases the specific heat approaches an asymptotic value at high temperatures, that can be given as

$$c_M = b/T^2 \qquad (2.37)$$

The specific heat at constant field, $c_H = (\partial E/\partial T)_H$, was evaluated in Section C for a mole of non-interacting, S = 1/2 paramagnetic ions. As a result c_H was found to be CH^2/T^2 in the high temperature limit, Eq. (2.15). By limiting the discussion to non-interacting S = 1/2 paramagnetic ions means that we assume U = 0. To find a more generally valid relation for c_H, one has to assume that there are interactions that cause U to be non zero, and then derive a corrected form of Eq. (2.15). In doing so, one obtains

$$c_H = (b + CH^2)/T^2, \tag{2.38}$$

if Eq. (2.37) is correct for c_M. Now the ratio c_M/c_H and thus also χ_{ad}/χ_T becomes

$$\chi_{ad}/\chi_T = c_M/c_H = b/(b + CH^2) \tag{2.39}$$

This expression shows several rather characteristic features, although it is strictly valid only under a number of assumptions. These include such restrictions as that the specific heats may be described by their asymptotic T^{-2} dependence, and that the magnetic field not be too strong ($g\mu_B H < 2kT$) so that saturation effects on χ_T can be omitted. Let us now consider the two interesting limits of $H \to 0$ and $H \to \infty$.

1. $\underline{H \to 0.}$ From Eq. (2.39) one immediately concludes that χ_{ad}/χ_T becomes 1, or $\chi_{ad} = \chi_T$. This means that relaxation effects can be neglected as the frequency dependent term in Eq. (2.35) is proportional to $\chi_T - \chi_{ad}$. One realizes this limit experimentally as H^2 in Eq. (2.39) becomes small compared to b/C. In fact $(b/C)^{\frac{1}{2}}$ is often called an internal field, and in cases where the external field H during the experiments is much weaker than $(b/C)^{\frac{1}{2}}$, one need not worry about spin-lattice relaxation, as $\chi_T = \chi_{ad}$.

2. $\underline{H \to \infty.}$ In strong external magnetic fields χ_{ad}/χ_T becomes zero. By strong fields we mean in this case $H \gg (b/C)^{\frac{1}{2}}$ a situation which can be realized even in fields that are far from being so strong that one has to consider saturation effects on χ_T (Chapter I-C). Under such circumstances the differential susceptibility may have any value between χ_T and 0 and one should be careful in interpreting experimental results.

The above comments suggest that apart from the relaxation time constant τ, the internal field $\sqrt{(b/C)}$ is also an important quantity in deciding whether the differential susceptibility yields χ_T or not.

The value of the internal field can be calculated from the specific heat at constant magnetization. Another possibility is to measure the high frequency susceptibility as a function of field. A plot of χ_T/χ_{ad} versus H^2 yields b/C, and in some cases this internal field is used to derive the high temperature approximation of the specific heat at constant magnetization. In fact this method is used extensively by Gorter and his co-workers in Leiden (22). A modern analogue of this way of measuring

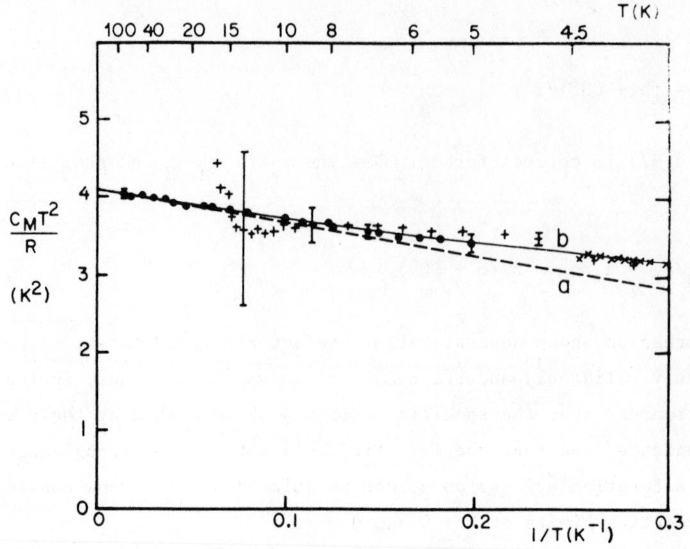

FIGURE 2.11 *High temperature magnetic specific heat for $Gd(OH)_3$ from high-frequency susceptibility measurements (•) and calorimetric measurements (+,×). There is good agreement between the two sets of data in the region of overlap, but owing to the relative error limits, only the high frequency data can be used to obtain the correct asymptotic fit to c_M. From Ref. 23.*

specific heats was developed by Skjeltorp and Wolf (23), a typical result being given in Figure 2.11, where $c_M T^2/R$ is plotted against $1/T$. This figure should show a constant value for this quantity if Eq. (2.37) is valid; in this case the appropriate corrections to Eq. (2.37) were obtained from the high frequency data.

Most of the phenomena discussed above do not change essentially if the system is not an ideal paramagnet. It would be beyond the scope of this book to describe all the detailed models. In the study of magnetic ordering by means of differential susceptibilities one should be aware of differences between χ_T and χ_{ad} similar to those described above. The behavior of χ_{ad}/χ_T as a function of field or temperature is less predictable in such cases, but it has been demonstrated that the thermodynamic relations for the adiabatic and isothermal susceptibility are correct (24).

REFERENCES

1. General references for this chapter include A.H. Morrish, "Physical Principles of Magnetism," J. Wiley and Sons, New York, 1965; J.C. Verstelle and D.A. Curtis, Handbuch d. Physik, S. Flügge, ed., Springer (Berlin, 1968), Vol. XVIII, pt. 1, p. 1; D. de Klerk, Handbuch d. Physik, Vol. XV, p. 38.
2. D.G. Kapadnis and R. Hartmans, Physica 22, 173 (1956).
3. J.W. Stout and E. Catalano, J. Chem. Phys. 23, 2013 (1955); see also W.O.J. Boo and J.W. Stout, J. Chem. Phys. 65, 3929 (1976).
4. P. Bloembergen and A.R. Miedema, Physica 75, 205 (1974).
5. K. Kopinga, T. de Neef, and W.J.M. de Jonge, Phys. Rev. B11, 2364 (1975).
6. W.J.M. de Jonge, C.H.W. Swüste, K. Kopinga, and K. Takeda, Phys. Rev. B 12 5858 (1975). See also K. Kopinga, P. van der Leeden, and W.J.M. de Jonge, Phys. Rev. B 14, 1519 (1976).
7. W.F. Giauque and D.P. MacDougall, J. Am. Chem. Soc. 57, 1175 (1935).
8. L.C. Hebel and C.P. Slichter, Phys. Rev. 113, 1504 (1959).
9. See, for example, Chapter 2 by R. Orbach and H.J. Stapleton in S. Geschwind, ed., "Electron Paramagnetic Resonance," Plenum Press, New York, 1972.
10. J.H. van Vleck, Phys. Rev. 57, 426 (1940).
11. J.M. Ziman, "Electrons and Phonons," Oxford University Press, London, 1960.
12. L. Cianchi and M. Mancini, Revista del Nuovo Cimento 2, 25 (1972).
13. See, for example, C. van Dijk, Thesis, Leiden, 1970.
14. A.C. de Vroomen, E.E. Lijphart and N.J. Poulis, Physica 47, 458 (1970).
15. J. Soeteman, L. Bevaart and A.J. van Duyneveldt, Physica 74, 126 (1974).
16. C.B.P. Finn, R. Orbach and W.P. Wolf, Proc. Phys. Soc. 77, 261 (1961).
17. C.L.M. Pouw and A.J. van Duyneveldt, Physica 83B, 163 (1976).
18. C.J. Gorter and A.J. van Duyneveldt, Proc. Conf. Low Temperature Physics, LT 13, Vol. 2, 621 (1973).
19. K.J. Standley and R.A. Vaughan, "Electron Spin Relaxation Phenomena in Solids," Adam Hilger, London, 1969.
20. R.A. Vaughan, Magn. Res. Reviews 4, 25 (1976).
21. H.B.G. Casimir and F.K. Du Pré, Physica 5, 507 (1938).
22. See, for example, C.J. Gorter, "Paramagnetic Relaxation," Elsevier, Amsterdam, 1947.
23. A.T. Skjeltorp and W.P. Wolf, Phys. Rev. B8, 215 (1973).
24. A.J. van Duyneveldt, J. Soeteman and L.J. de Jongh, J. Phys. Chem. Solids 36, 481 (1975).

CHAPTER III

PARAMAGNETISM: ZERO-FIELD SPLITTINGS

A. INTRODUCTION

The subject of zero-field splittings, introduced briefly by Figure 1.6b, requires a far more detailed description, for it is central to much that is of interest in magnetism. Zero-field splittings are often responsible for deviations from Curie law behavior, give rise to a characteristic specific heat behavior, and limit the usefulness of certain substances for adiabatic demagnetization. Furthermore, zero-field splittings cause a single-ion anisotropy which is important in characterizing anisotropic exchange, and are also one of the sources of canting or weak ferromagnetism (Chapt. VII).

The situation in Figure 1.6 is repeated in Figure 3.1 where for convenience we set $L = 0$ for an $S = 3/2$ system. In an axial crystalline field, the fourfold degeneracy, $m = \pm 1/2, \pm 3/2$, is partially resolved with the $\pm 3/2$ states separated an amount $2D$ (in units of energy) from the $\pm 1/2$ states. The population of the levels then depends, by the Boltzmann principle, on the relative values of the parameter D and the thermal energy, kT. Of large consequence for many magnetic systems is the fact that D not only has magnitude but also sign. As drawn, $D > 0$; if $D < 0$, then the $\pm 3/2$ levels have the lower energy.

The spin-Hamiltonian approach which is used in electron paramagnetic resonance (EPR) studies is also valuable here. The Zeeman operator used extensively in Chapter I may simply be written as

$$\mathcal{H} = g\mu_B H_z S_z \tag{3.1}$$

and since states labeled by their value of m are eigenstates of the

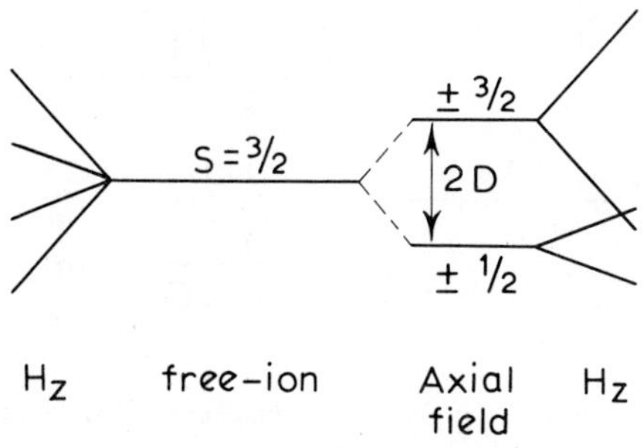

FIGURE 3.1 *The behavior of an S = 3/2 state (center). The usual Zeeman splitting is illustrated on the left, while on the right the effect of zero-field splitting is illustrated. The degeneracies are partially resolved before the magnetic field affects the levels.*

operator S_z, the splittings proportional to a field are obtained immediately. In the present case of an axial distortion, addition of the term

$$\mathcal{H}' = D\,[\,S_z^{\,2} - \tfrac{1}{3}S(S + 1)\,] \tag{3.2}$$

where S_z is again a spin operator and S is the spin will reproduce the splitting diagram of Figure 3.1 when the magnetic field and principal axis are parallel.

It is sometimes necessary to add a rhombic term of the form

$$\mathcal{H}'' = E(S_x^{\,2} - S_y^{\,2}) \tag{3.3}$$

to the spin-Hamiltonian, in order to take account of the anisotropy in the plane perpendicular to the principal axis. This term has the effect of increasing the zero-field splitting or splitting further certain degeneracies. It is conventional (1) to choose the axes such that D is always larger than E.

The name "zero-field splitting" arises from the fact that the splitting occurs in the absence of a magnetic field. As we learn from crystal field theory, it can be ascribed to the electrostatic field of the ligands. The effect of zero-field splittings on the magnetic energy levels and resulting magnetic properties when a field is applied will be explored below.

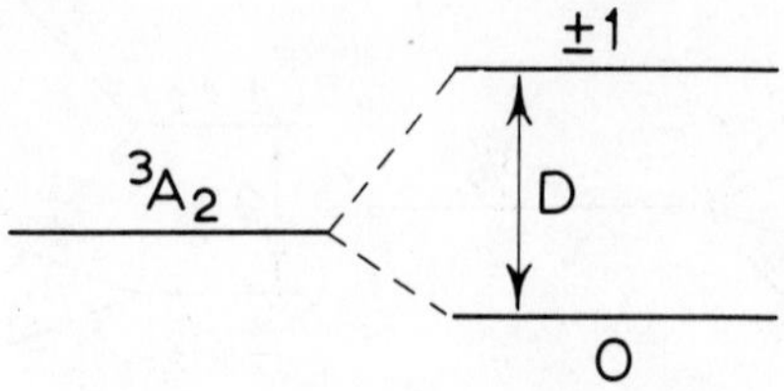

FIGURE 3.2 *Zero-field splitting (positive) of a nickel(II) (or vanadium(III)) ion.*

B. SCHOTTKY ANOMALIES

One of the most important features of a zero-field splitting is the broad maximum that it causes in the specific heat, which for historical reasons is called a Schottky anomaly. Consider the spin-1 nickel ion with $^3A_2$ ground state (2) with positive parameter D as illustrated in Figure 3.2. Application of the spin-Hamiltonian, Eq. (3.2), gives rise to a doubly-degenerate level with energy D above a non-degenerate level; for convenience alone, we ignore the rhombic term. The single-ion partition function (Chapt. I-C) is

$$Z = 1 + 2e^{-D/kT} \tag{3.4}$$

and straightforward application of the thermodynamic relation

$$C = \frac{\partial}{\partial T}\left(RT^2 \frac{\partial \ln Z}{\partial T}\right) \tag{3.5}$$

leads immediately to the magnetic specific heat,

$$C = \frac{2R(D/kT)^2 \exp(-D/kT)}{[1 + 2\exp(-D/kT)]^2} \tag{3.6}$$

which has the shape illustrated in Figure 3.3. This curve has a maximum at approximately $T_{max} = 0.4$ D/k, where D/k is the zero-field splitting expressed in Kelvins. It is essential to recall that this specific heat is the magnetic contribution alone, which is of course superimposed on the lattice heat capacity. As an example, Figure 3.4 illustrates the specific heat of α-$NiSO_4 \cdot 6H_2O$ (3). A zero-field splitting of about 7 K provides the magnetic contribution, which falls on top of the T^3 lattice term. Clearly, the smaller the zero-field splitting, the lower the temperature

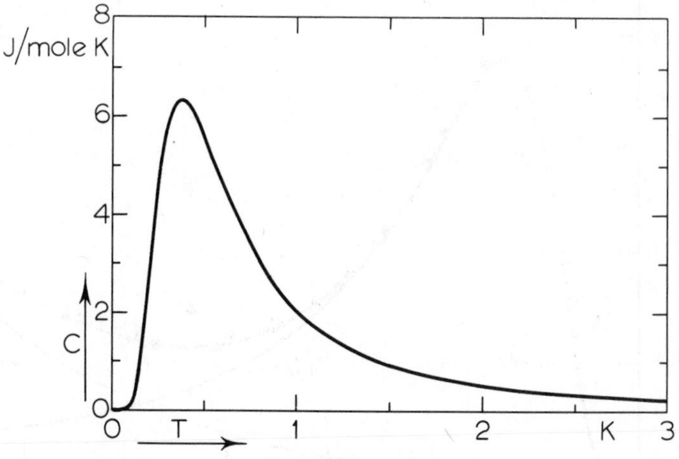

FIGURE 3.3 *Schottky specific heat for a* $^3A_2$ *ground state for* $D/k = 1$ *K.*

at which it will give rise to a maximum, which can in turn be measured more accurately.

As an example of this situation, the magnetic specific heat of $V(urea)_6Br_3.3H_2O$ is shown in Figure 3.5, together with the fit (4) to the Schottky curve, Eq. (3.6). The lattice contribution was evaluated by the corresponding states procedure, making use of the specific heat of the isomorphic $Fe(urea)_6Cl_3.3H_2O$. Although this is not a diamagnetic compound, the measured specific heat above 1 K gave no evidence for any magnetic effects, as one would expect because the zero-field splitting of the ^6S ground state of this compound is much smaller than kT in this region.

More generally, consider a level of degeneracy g_1 which is δ in energy above a level of degeneracy g_o. (It is of course not necessary to restrict the analysis to a two level system.) Then, following the procedure described above,

$$C = \frac{R(\delta/kT)^2 (g_o/g_1)\exp(\delta/kT)}{[1 + (g_o/g_1)\exp(\delta/kT)]^2} \tag{3.7}$$

Eq. (3.7) is plotted in Figure 3.6 with reduced parameters for several values of g_o and g_1. The height of the maximum depends only on the relative values of g_o and g_1, while the parameter δ determines the

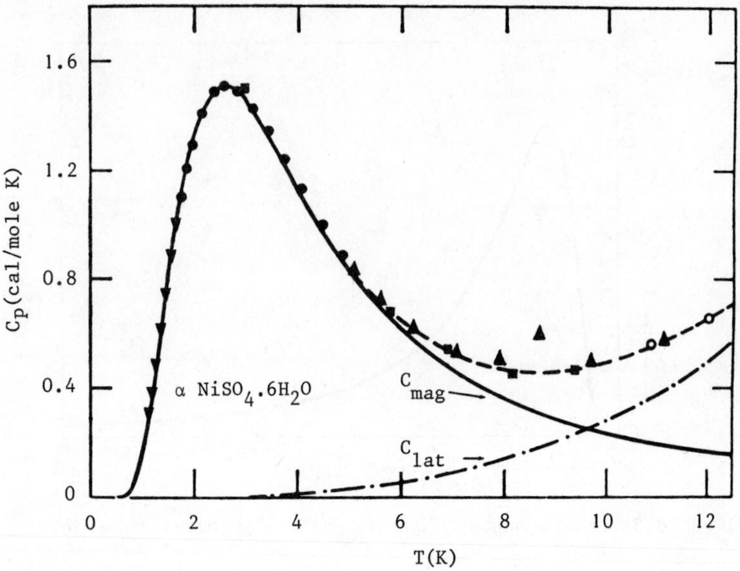

FIGURE 3.4 *Heat capacity of* α - $NiSiO_4.6H_2O$. *From ref. 3.*

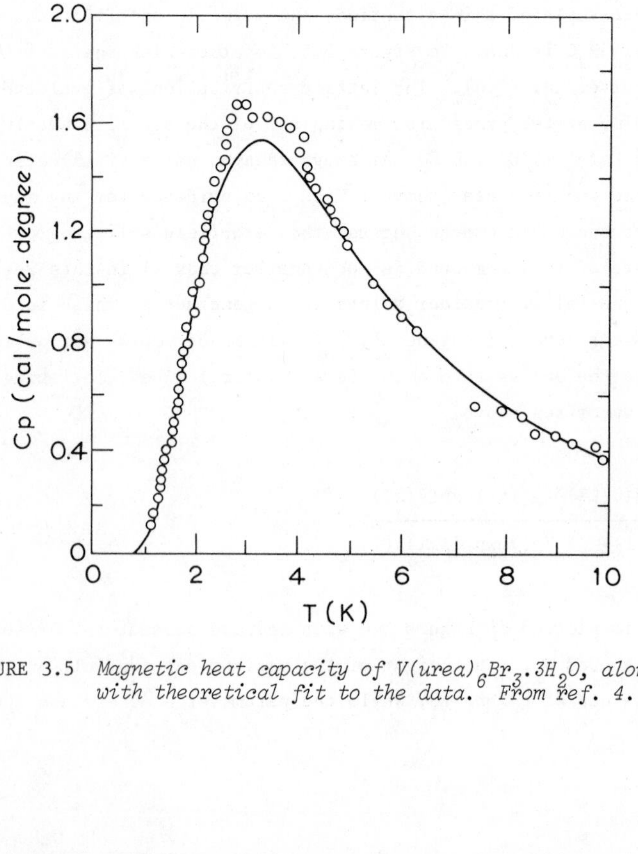

FIGURE 3.5 *Magnetic heat capacity of* $V(urea)_6Br_3.3H_2O$, *along with theoretical fit to the data. From ref. 4.*

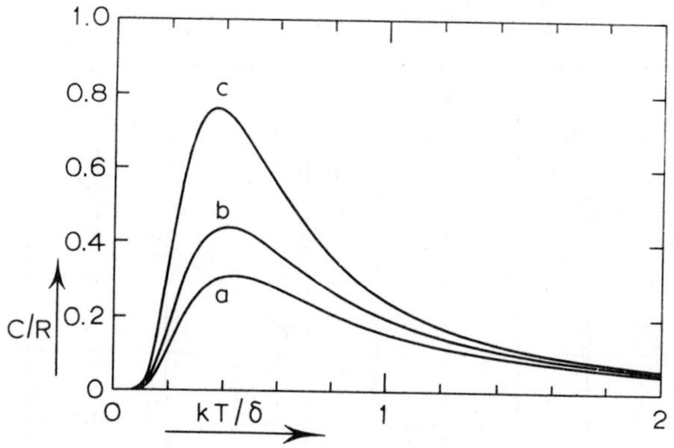

FIGURE 3.6 *Schottky heat capacity of a two-level system for several values of the relative degeneracy,* g_1/g_0. *Curve a:* $g_1/g_0 = \frac{1}{2}$, *b:* $g_1/g_0 = 1$, $g_1/g_0 = 2$.

position of the maximum on the T axis.

It is common to find substances, particularly of Cr(III), Mn(II) and Fe(III), with δ of the order of only 0.1 K, a situation that is more accurately studied by means of EPR rather than by specific heat measurements. At those temperatures for which $\delta \ll kT$, the exponentials in Eq. (3.7) may be expanded to yield

$$C = \frac{R(g_o/g_1)}{(1 + g_o/g_1)^2}(\delta/kT)^2 = \frac{\text{const}}{T^2} . \tag{3.8}$$

That is, the high-temperature tail of the Schottky specific heat varies as T^{-2}. This result holds at higher temperatures not only for a two-level system, but for a system with any number of closely-spaced levels. Magnetic ions can interact (Chapt. V) by both magnetic-dipole-magnetic-dipole and exchange interactions, and these interactions also cause a specific heat maximum which varies as T^{-2} at high temperatures. Furthermore nuclear spin-electron spin hyperfine interactions become important at very low temperatures, and can cause a Schottky-like contribution. All these possible contributions can be included in one parameter if we write the high - temperature magnetic specific heat as

$$C/R = \frac{A/R}{T^2} \tag{3.9}$$

and A/R is a (high-temperature) measure of the importance of the inter-

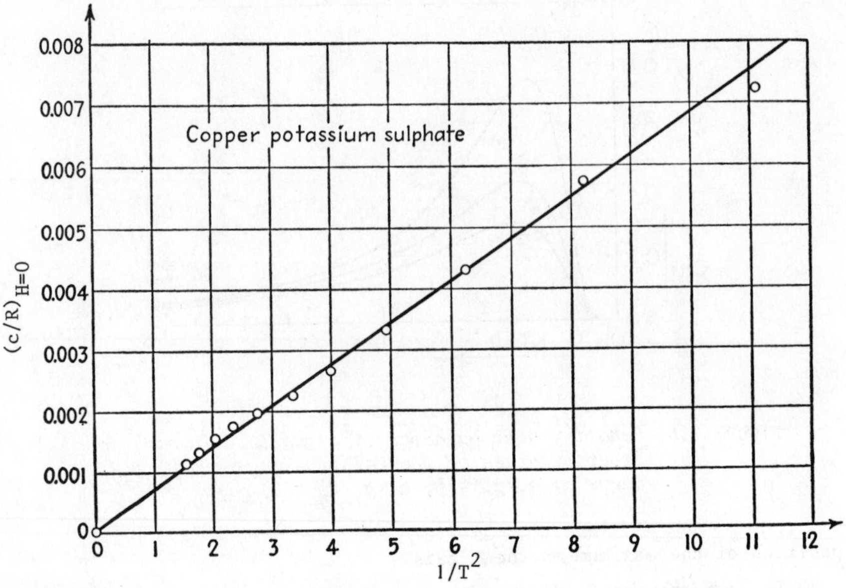

FIGURE 3.7 T^{-2} *dependence of C/R for copper potassium sulfate.*

actions combined with the resolution of the energy levels. Eq. (3.9) is applied to $K_2Cu(SO_4)_2 \cdot 6H_2O$ in Figure 3.7. Some typical values of A/R for salts which have been discussed so far are:

chrome alum	$A/R = 0.018\ K^2$
iron alum	$0.013\ K^2$
$Gd_2(SO_4)_2 \cdot 8H_2O$	$0.35\ K^2$
$Ce_2Mg_3(NO_3)_{12} \cdot 24H_2O$ (CMN)	$6.1 \times 10^{-6}\ K^2$

The very small value of A/R for CMN indicates that all the factors that split the lowest energy level are also very small.

C. ADIABATIC DEMAGNETIZATION

We return to the subject of adiabatic demagnetization through the relation for the entropy of a magnetic system (5),

$$S = RT\left(\frac{\partial \ln Z}{\partial T}\right)_H + R\ln Z \qquad (3.10)$$

This relation can be written down completely when Z is known, which as a

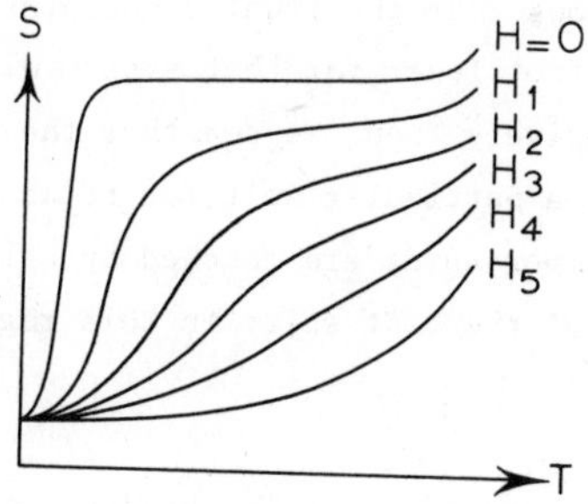

FIGURE 3.8

Entropy of a magnetic ion subsystem as a function of temperature for several values of the magnetic field. In the region where S(H=0) is constant, the energy U is a constant, and thus $c_M = 0$.

practical matter is true only when there are no interactions between ions. If only a zero-field splitting δ is considered the entropy behaves simply in the limits

1) $kT \gg \delta, \quad S = S_1(H/T)$
2) $kT \leq \delta, \quad S = S_2(H,T)$

Several curves are displayed in Figure 3.8 for various values of the field H.

The requirement of finding a good refrigerant crystal for adiabatic demagnetization work was the initial impetus for measuring zero-field splittings. Consider a two level system. The smaller the splitting, the more the populations of the two levels will be equalized. This corresponds to more disorder or larger entropy of the spin-system, and so, as illustrated in Figure 3.9, the entropy curve at zero field for the system with smaller zero-field splitting lies above that for the system with the larger zero-field splitting. Application of the adiabatic demagnetization

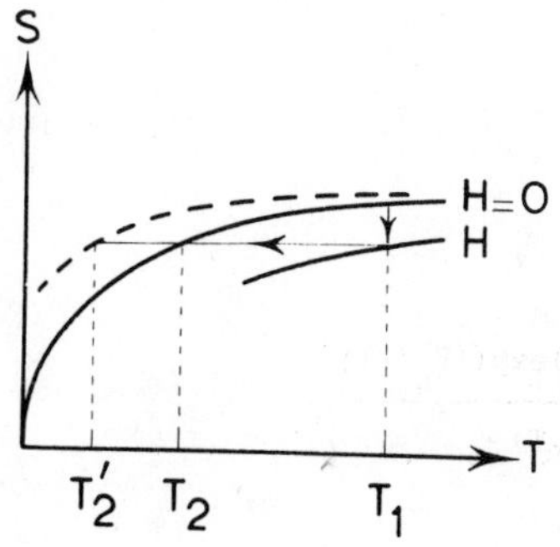

FIGURE 3.9

Effect of zero-field splitting on the final temperature.

cycle, as was illustrated in Chapt. II, shows that the final temperature after the adiabatic demagnetization is in fact lower for that system with the smaller zero-field splitting. Qualitatively then, we see that the quantity A is a guide to the usefulness of a particular salt for adiabatic demagnetization, and in fact the lowest temperatures are reached by salts in which A is small. Clearly, CMN is one of the best salts in this regard.

D. VAN VLECK'S EQUATION

Before we can calculate the effect of zero-field splittings on magnetic susceptibilities, we require a more general method for calculating these susceptibilities. If the energy levels of a system are known, the magnetic susceptibility may always be calculated by application of Van Vleck's equation. The standard derivation (6) follows, along with several typical applications.

Let the energy, E_n, of a level be developed in a series in the applied field: $E_n = E_n^o + HE_n^{(1)} + H^2E_n^{(2)} + \ldots..$, where, in the standard nomenclature, the term linear in H is called the first-order Zeeman term, and the term in H^2 is the second order Zeeman term. Since $\mu_n = -\partial E_n/\partial H$, the total magnetic moment, M, for the system follows as before:

$$M = \frac{N \sum_n \mu_n \exp(-E_n/kT)}{\sum_n \exp(-E_n/kT)}$$

Now,

$$\exp(-E_n/kT) = \exp\{-(E_n^o + HE_n^{(1)} + \ldots..)/kT\} \approx (1 - HE_n^{(1)}/kT)\exp(-E_n^o/kT)$$

by expansion of the exponential, and

$$\mu_n = -\frac{\partial E_n}{\partial H} = -E_n^{(1)} - 2HE_n^{(2)} + \ldots$$

To this approximation we obtain

$$M = N\frac{\sum_n(-E_n^{(1)} - 2HE_n^{(2)})(1 - HE_n^{(1)}/kT)\exp(-E_n^o/kT)}{\sum_n \exp(E_n^o/kT)(1 - HE_n^{(1)}/kT)}$$

We limit the derivation to paramagnetic substances, as distinct from ferromagnetic ones so that the absence of permanent polarization in zero magnetic field (i.e., M = 0 at H = 0) requires that

$$\sum_n -E_n^{(1)} \exp(-E_n^o/kT) = 0$$

Retaining only terms linear in H,

$$M = N \frac{H \sum_n [(E_n^{(1)})^2/kT - 2E_n^{(2)}] \exp(-E_n^o/kT)}{\sum_n \exp(-E_n^o/kT)}$$

Since the static susceptibility is $\chi = M/H$, the final result is:

$$\chi = \frac{N \sum_n [(E_n^{(1)})^2/kT - 2E_n^{(2)}] \exp(-E_n^o/kT)}{\sum_n \exp(-E_n^o/kT)} \tag{3.11}$$

The degeneracy of any of the levels has been neglected here, but of course the r-degeneracy of any level must be summed r times.

The general form of the spin-only susceptibility is obtained as follows. Consider an orbital singlet with 2S + 1 spin degeneracy. The energy levels are at $mg\mu_B H$, where m spans the values from +S to -S. Note that the energy levels correspond to

$$E_n^o = E_n^{(2)} = 0, \quad E_n^{(1)} = mg\mu_B$$

since the zero of energy can be taken as that of the level of lowest energy in the magnetic field. Then, applying these results to Eq. (3.11), i.e.,

$$\chi = \frac{Ng^2\mu_B^2}{kT} \frac{(-S)^2 + (-S+1)^2 + \ldots\ldots + (+S)^2}{2S + 1}$$

$$\chi = Ng^2\mu_B^2 S(S + 1)/3kT \tag{3.12}$$

since

$$\sum_{-S}^{S} m^2 = \tfrac{1}{3} S(S + 1)(2S + 1).$$

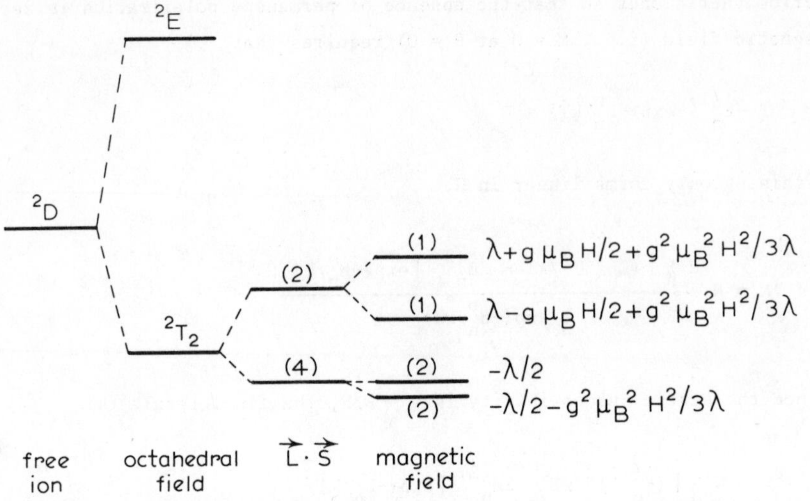

FIGURE 3.10 *Crystal field, spin-orbit, and magnetic field splitting of the ^2D energy level of titanium(III).*

This result was presented earlier as Eq. (1.6).

Kotani (7) applied Van Vleck's equation to octahedral complexes of trivalent titanium. The Hamiltonian for the d^1 ion perturbed by both a magnetic field and spin-orbit coupling is:

$$\mathcal{H} = \lambda\, \vec{L}.\vec{S} + \mu_B(\vec{L} + g\vec{S}).\, \vec{H} \tag{3.13}$$

The resulting energy level diagram is given in Figure 3.10. The numbers in parentheses indicate the degeneracy of a particular level. We ignore the contribution to paramagnetism from the ^2E states, for they are some 20,000 cm^{-1} higher in energy. (The contribution of any energy level n is proportional to $\exp(-E_n/kT)$, and $kT \approx 205$ cm^{-1} at room temperature.)

Therefore, applying Eq. (3.11) to the energy levels sketched in Figure 3.10,

$$\chi/N = [2(0^2/kT - 2\times 0)\exp(\lambda/2kT) + 2(0^2/kT + 2g^2\mu_B^2/2\lambda)\exp(\lambda/2kT) + 2(g^2\mu_B^2/4kT - 2g^2\mu_B^2/3\lambda)\exp(-\lambda/kT)]/2[\exp(\lambda/2kT) + \exp(\lambda/2kT) + \exp(-\lambda/kT)]$$

which reduces to

$$\chi = N g^2 \mu_B^2 \mu^2 / 3kT \tag{3.14}$$

with

$$\mu^2 = \frac{8 + (3\lambda/kT - 8)\exp(-3\lambda/2kT)}{4(\lambda/kT)\,[2 + \exp(-3\lambda/2kT)]} \tag{3.15}$$

Note that the Curie Law does not hold in this case. In fact, in the limit as $T \to \infty$, the exponentials may be expanded, the first term retained, and one finds $\mu^2 \to 5/4$ while $\mu^2 \to 0$ as $T \to 0$. How can a system with an unpaired electron have zero susceptibility at 0 K? The spin and orbital angular momenta cancel each other out. Note also that $\mu^2 \to 5/4$ as $\lambda \to 0$.

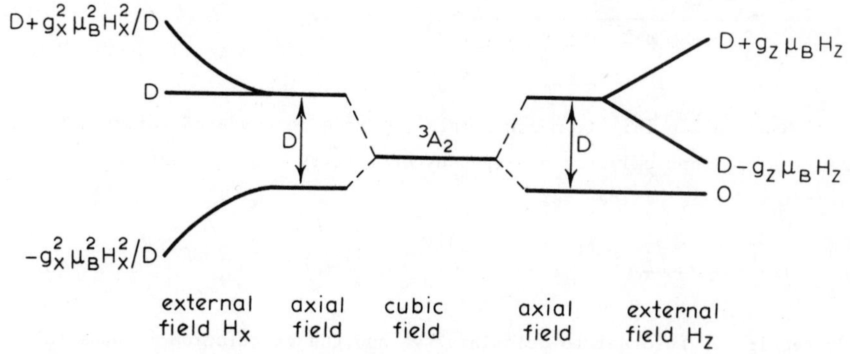

FIGURE 3.11 *Nickel(II) with an internal axial field and an external magnetic field.*

E. PARAMAGNETIC ANISOTROPY

Zero-field splittings are one of the most important sources of paramagnetic anisotropies. With the aid of the energy level diagram of nickel(II) in a weak tetragonal field, Figure 3.11, and Van Vleck's equation, we are now able to calculate the susceptibilities with the measuring field parallel or perpendicular to the principal axis. The parameter D measures the zero-field splitting of the ground state. Our procedure yields for the parallel susceptibility,

$$\chi_{\parallel}/N = \frac{(g_z\mu_B)^2}{kT} \frac{\exp(-D/kT) + 0\exp(-0/kT) + (g_z^2\mu_B^2/kT)\exp(-D/kT)}{[1 + 2\exp(-D/kT)]}$$

Assuming D may be large (i.e., the exponentials are not expanded),

$$\chi_{\parallel} = \frac{2Ng_z^2\mu_B^2}{kT} \frac{\exp(-D/kT)}{[1 + 2\exp(-D/kT)]} \qquad (3.16)$$

Frequently, however, $D \ll kT$ (with $T \approx 300$ K at room temperature, and D/k is often of the order of a few Kelvins, and rarely is as large as 25 K), and this reduces to

$$\chi_{\parallel} \approx \frac{2Ng_z^2\mu_B^2(1 - D/kT)}{kT(1 + 2 - 2D/kT)} \approx \frac{2Ng_z^2\mu_B^2}{3kT}(1 - D/3kT) \qquad (3.17)$$

Many susceptibility measurements are made on powdered paramagnetic samples, so that only the average susceptibility, $\langle\chi\rangle$, is obtained. The quantity $\langle\chi\rangle$ is defined as

$$\langle\chi\rangle = \frac{\chi_{\parallel} + 2\chi_{\perp}}{3} \qquad (3.18)$$

We require $\chi_{\perp}$ in order to calculate $\langle\chi\rangle$ and the calculation is done in the following fashion.

The set of energy levels in a field used above for the calculation of $\chi_{\parallel}$ is of course valid only for z as the axis of quantization. If a anisotropic single crystal is oriented such that the external field is normal to the principal molecular or crystal field axis, we need to consider the effect of this field H_x (H_y is equivalent in trigonal and tetragonal fields) on the energy levels. The Hamiltonian is

$$\mathcal{H}_x = g_x\mu_B H_x S_x = \frac{g_x\mu_B H_x}{2}(S_+ + S_-) \qquad (3.19)$$

where $S_\pm = S_x \pm iS_y$, and S_x, S_y are the operators for the x,y components of electron spin, respectively. The problem is to calculate the eigenvalues of this Hamiltonian for a system with energy levels labeled

$m = \pm 1$ in zero-field, which are D in energy above a level with $m = 0$. That is, we require the eigenvalues of the 3 x 3 matrix made up from the set of matrix elements $\langle m|\mathcal{H}_x|m'\rangle$. We make use of the standard formulas

$$\langle m|S_z|m\rangle = m$$
$$\langle m\pm 1|S_\pm|m\rangle = [S(S+1) - m(m\pm 1)]^{\frac{1}{2}}$$

and see that the diagonal elements of Eq. (3.19) are all zero. The only nonzero elements are

$$\langle \pm 1|\mathcal{H}_x|0\rangle = \langle 0|\mathcal{H}_x|\pm 1\rangle = g_x\mu_B H_x/\sqrt{2}$$

and the matrix takes the form

$$\begin{array}{ccc} +1 & -1 & 0 \end{array}$$
$$\begin{bmatrix} D & 0 & g_x\mu_B H_x/\sqrt{2} \\ 0 & D & g_x\mu_B H_x/\sqrt{2} \\ g_x\mu_B H_x/\sqrt{2} & g_x\mu_B H_x/\sqrt{2} & 0 \end{bmatrix}$$

Subtracting the eigenvalues W from each of the diagonal terms and setting the determinant of the matrix equal to zero results in the following cubic equation

$$W^3 + W^2(-2D) + W\,(D^2 - g_x^2\mu_B^2H_x^2) + g_x^2\mu_B^2H_x^2D = 0$$

which has roots

$$W_1 = 1/2\,[\,D - (D^2 + 4g_x^2\mu_B^2H_x^2)^{\frac{1}{2}}\,]$$
$$W_2 = D$$
$$W_3 = 1/2\,[\,D + (D^2 + 4g_x^2\mu_B^2H_x^2)^{\frac{1}{2}}\,]$$

Making use of the expression $(1+x)^{\frac{1}{2}} \approx 1 + (1/2)x$, which corresponds to small magnetic fields, the energy levels become

$$W_1 = -g_x^2\mu_B^2H_x^2/D$$
$$W_2 = D \qquad (3.20)$$
$$W_3 = D + g_x^2\mu_B^2H_x^2/D.$$

These energy levels are plotted, schematically, as a function of external

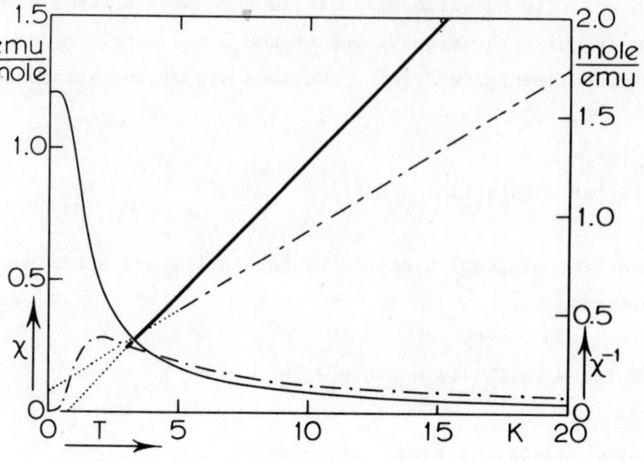

FIGURE 3.12 *Parallel and perpendicular susceptibilities, and their inverses, for a (hypothetical) nickel(II) ion with* $g_{\parallel} = g_{\perp} = 2.2$ *and* $D/k = 3$ K; *drawn lines:* $\chi_{\perp}$ *and* $\chi_{\perp}^{-1}$, *dotted-drawn lines:* $\chi_{\parallel}$ *and* $\chi_{\parallel}^{-1}$.

field in Figure 3.11. They are readily inserted into Van Vleck's equation to yield

$$\chi_x/N = \frac{2g_x^2\mu_B^2/D - (2g_x^2\mu_B^2/D)\exp(-D/kT)}{[1 + 2\exp(-D/kT)]}$$

or

$$\chi_x = \chi_{\perp} = \frac{Ng_{\perp}^2\mu_B^2}{3kT}\frac{6kT}{D}\left[\frac{1 - \exp(-D/kT)}{1 + 2\exp(-D/kT)}\right] \qquad (3.21)$$

Neglecting the difference between $g_{\parallel}$ and $g_{\perp}$, we also obtain, after averaging,

$$\langle\chi\rangle = 2\frac{Ng^2\mu_B^2}{3kT}\frac{[2/x - 2\exp(-x)/x + \exp(-x)]}{[1 + 2\exp(-x)]}$$

where $x = D/kT$.

Although it was assumed throughout the derivation that $D > 0$, the equations are equally valid if $D < 0$. The complete solutions for $\chi_{\parallel}$ and $\chi_{\perp}$ are plotted in Figure 3.12 for a typical set of parameters. It will be noticed that Curie-Weiss-like behavior is found at temperatures high with respect to D/k, but marked deviations occur at $D/kT \leq 1$. The perpendicular susceptibility approaches a constant value at low temperatures, while the parallel susceptibility goes to zero. A measurement of $\langle\chi\rangle$ alone in this temperature region would offer few clues as to the situation.

Plots of χ^{-1} vs. T are also seen to be linear at high temperatures and to deviate when $D/kT \rightarrow 1$. If the straight-line portion is extrapolated to $\chi^{-1} = 0$, non-zero intercepts are obtained. This behavior is often treated empirically by means of the Curie-Weiss law, where χ is written

$$\chi = \frac{C}{T-\theta} \tag{3.22}$$

where 1/C is the slope of the curve (and C is the supposed Curie constant) and θ is the intercept of the curve with the T axis (positive or negative). This result illustrates merely one of the sources of Curie-Weiss behavior, exchange being another, and shows that caution is required in ascribing the non-zero value of θ to any one particular source.

The ground-state energy-levels of Figure 3.2 are applicable particularly to vanadium(III) and nickel(II). The parameter D/k is relatively larger for vanadium, often being of the order of 5 to 10 K (4). Interestingly, D > 0 for every case investigated to date. Nickel usually has a smaller D/k, some 0.1 to 6 K, and both signs of D have been observed. There seems to be no rational explanation for these observations at present. As will be discussed in Chapt. VIII, there are several compounds of nickel studied recently that seem to have very large zero-field splittings, so the observations described here may no longer be valid when a larger number and variety of compounds have been studied. A typical set of data is illustrated in Figure 3.13; only two parameters, the g values and D, are required in order to fit the measurements on $[C(NH_2)_3]V(SO_4)_2 \cdot 6H_2O$ (8). On the other hand, only one parameter is required in the analysis of Schottky anomalies in terms of zero-field splitting. The measurement and interpretation of paramagnetic anisotropies have been reviewed recently (9).

F. EFFECTIVE SPIN

This is a convenient place to introduce the concept of the effective spin of a metal ion, which is not necessarily the same as the true spin.

The two assignments of spin are the same for a copper(II) ion. This ion has nine d electrons outside the argon core and so has a spin-1/2

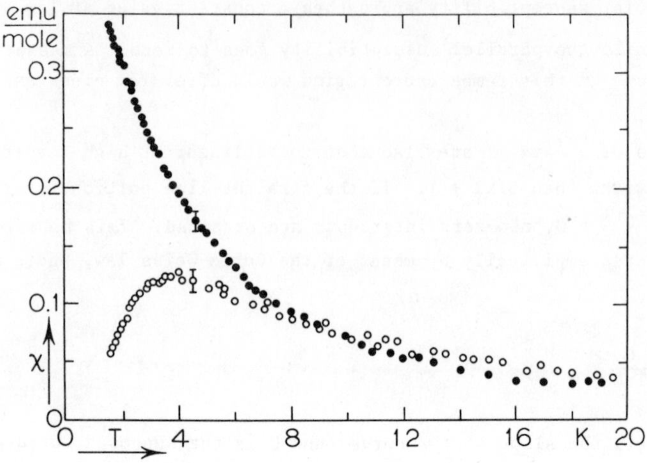

FIGURE 3.13 *Parallel and perpendicular susceptibilities of* $[C(NH_2)_3]V(SO_4)_2 \cdot 6H_2O$. *From ref. 8.*

configuration no matter what geometry the ion is placed in, nor does it depend on the strength of the crystal or ligand field. Thus, it has a spin (Kramers) doublet as the ground state, which is well-isolated from the optical states. One consequence of this is that EPR spectra of Cu(II) may usually be obtained in any lattice (i.e., liquid or solid) at any temperature.

Consider divalent cobalt, however, which is $3d^7$. With a $^4T_1$ ground state in an octahedral field, there are three unpaired electrons, and numerous Co(II) compounds exhibit spin-3/2 magnetism at 77 K and above (10). However, the spin-orbit coupling constant of Co(II) is large ($\lambda \sim -180$ cm^{-1} for the free ion), and so the true situation here is better described by the energy level diagram in Figure 3.14. It will be seen there that, under spin-orbit coupling, the $^4T_1$ state splits into a set of three levels, with the degeneracies as noted on the figure in parentheses. At elevated temperatures, the excited states are occupied and the spin-3/2 configuration, with an important orbital contribution, obtains. Fast electronic relaxation also occurs, and no EPR is observed. At low temperatures - say, below 20 K - only the ground state is occupied, but this is a doubly-degenerate level, and even though this is a spin-orbit doublet we may characterize the situation with an effective spin, S' = 1/2 This is consistent with all available EPR data, as well as with magnetic susceptibility results. One further consequence of this situation is

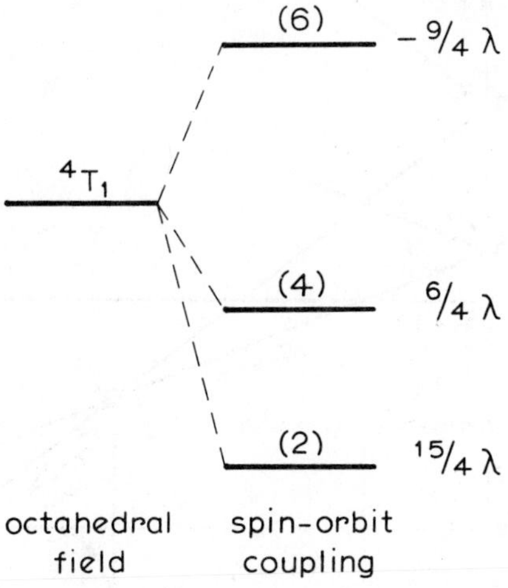

FIGURE 3.14 *Fine-structure splitting of the lowest levels of cobalt(II) under the action of the combined octahedral crystalline field, spin-orbit interaction, and a magnetic field. Degeneracies are given in parentheses: the spin-orbit coupling constant, λ, is negative for cobalt(II).*

that the effective, or measured, g values of this level deviate appreciably from the free spin value of 2; in fact (11) for octahedral cobalt(II) the three g-values are so unusual that they sum to 13!

Let us return to our now-familiar example of nickel(II), Figure 3.2, but change the sign of D to a negative value and thereby invert the figure. Note that if $|D| >> kT$, only the double-degenerate ground state will be populated, a spin doublet obtains, and $S' = \frac{1}{2}$. The g-values will be quite anisotropic for, using primes to indicate effective g values, $g_{\parallel}' = 2g_{\parallel}$ and $g_{\perp}' = 0$. The calculation differs from the cobalt situation in that the spin-orbit coupling is not as important. These g-values may be obtained by considering the calculation which was performed above in Section E.

Similar situations occur with tetrahedral Co^{2+} in both Cs_3CoCl_5 (12) and Cs_2CoCl_4 (13) and with Ni^{2+} in tetrahedral $NiCl_4{}^{2-}$ (14).

G. DIRECT MEASUREMENT OF D

There has recently (15) been a direct measurement of D in $FeSiF_6 \cdot 6H_2O$, a trigonal crystal which is easy to orient. The technique employed illustrates some of the results that are becoming available

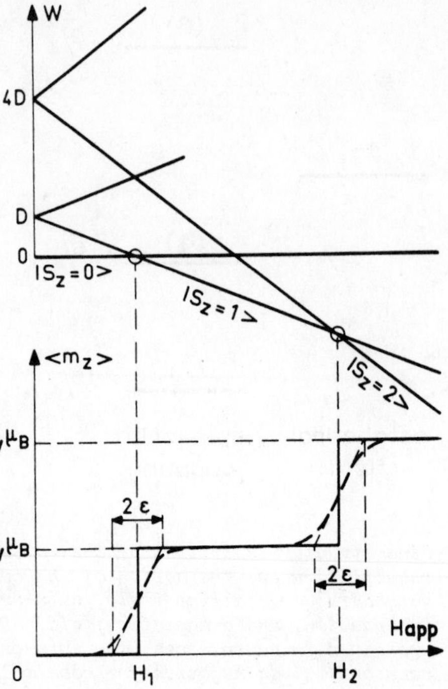

FIGURE 3.15 *Electronic level scheme and calculated magnetization at 0 K (full line) and finite temperature (dashed line) of Fe(II) in a magnetic field along the z axis, D >> λ.*

through the use of high-field magnets.

The ferrous ion, Fe^{2+}, has a $3d^6$ electronic configuration, which gives rise to a ^5D ground state. The energy level scheme, illustrated in Figure 3.15, has two doubly degenerate levels respectively D and 4D in energy above the non-degenerate ground state. Application of a magnetic field, as illustrated, causes the usual Zeeman interaction, and it will be noticed that a doubly-degenerate ground state occurs at each of the fields labeled H_1 and H_2. There can be no magnetization at $T << D/k$ when the external field, H_{ext}, is less than H_1, but a measurement of M when $H_{ext} = H_1$ or H_2 should be non-zero. The large fields required are necessarily pulsed fields, which in turn requires a rapid measurement of M; one must also be careful about spin-lattice relaxation effects in such experiments.

A quick glance at Figure 3.15 suggests that $H_1 = g_{\parallel}\mu_B D$ and $H_2 = 3g_{\parallel}\mu_B D$. Assuming $g_{\parallel} = 2$, an increase in M and therefore a cross-over was found at H_1 = 132 kOe. The derived value of D, 12.2 $\pm$ 0.2 cm^{-1}, may be compared to that obtained from susceptibility measurements (10.5 - 10.9 cm^{-1}). The field H_2 is then calculated at about 400 kOe, and indeed was measured as 410 kOe. Similar measurements have recently been reported on $[Ni(C_5H_5NO)_6](ClO_4)_2$ and $[C(NH_2)_3]V(SO_4)_2 \cdot 6H_2O$ (16).

H. ELECTRON PARAMAGNETIC RESONANCE (EPR)

Electron paramagnetic resonance, also called electron spin resonance, will not be treated in detail in this book, for there are so many other excellent sources available (17). Nevertheless, EPR is so implicitly tied up with magnetism that it is important just for the sake of completeness to include this short section on the determination of crystal field splittings by EPR. While the magnetic measurements that are the major topic of this book measure a property that is thermally averaged over a set of energy levels, EPR measures properties of the levels individually.

Thus, return to Figure 1.2, where we see that two levels are separated by an energy $\Delta E = g_{\parallel}\mu_B H_z$. Following the Planck relationship, if we set

$$\Delta E = h\nu = g_{\parallel}\mu_B H_z \tag{3.23}$$

we then have the basic equation of paramagnetic resonance spectroscopy. The quantities h and μ_B are fundamental constants, $g_{\parallel}$ is a constant for a particular orientation of a given substance, and so we see that the basic experiment of EPR is to measure the magnetic field H_z at which radiation of frequency ν is absorbed. For ease of experiment, the usual procedure is to apply radiation of a constant frequency and then the magnetic field is scanned. For a substance with g = 2, the common experiment is done at X-band, with a frequency of some 9 GHz, and absorption occurs at approximately 3400 gauss. The energy separation involved is approximately 0.3 cm^{-1}. Q-band experiments, at some 35 GHz are also frequently carried out, with the energy separation that can be measured then of the order of 1 cm^{-1}.

As we have seen, zero-field splittings or other effects can sometimes cause energy levels to be separated by energies larger than these relatively small values. In those cases, EPR absorption cannot take place between those energy levels. If that particular electronic transition should be the only allowed transition, then there will be no EPR spectrum at all, even in an apparently normal paramagnetic substance. This of course can happen with non-Kramers ions.

The usual selection rule for electron resonance is $\Delta m = \pm 1$, although a variety of forbidden transitions are frequently observed in certain situations. Paramagnetic ions are usually investigated as dopants in diamagnetic host lattices. This procedure minimizes dipole-dipole effects, and so sharper lines are thereby observed. Exchange effects upon the spectra are also minimized by this procedure. Since paramagnetic relaxation is temperature dependent a decrease in temperature lengthens the relaxation times and thereby also sharpens the lines.

Now, return to Figure 3.1, the energy level diagram for an $S = 3/2$ ion with and without zero-field splitting. Because of the $\Delta m = \pm 1$ selection rule, only one triply-degenerate line is observed at $h\nu = g_{\parallel}\mu_B H_z$ when the zero-field splitting is identically zero.

But, when the zero-field splitting is $2|D|$, the $+1/2 \leftrightarrow -1/2$ transition remains at $h\nu = g_{\parallel}\mu_B H_z$, and the $+3/2 \leftrightarrow -3/2$ transition is formally forbidden. The transitions $+1/2 \leftrightarrow +3/2$ and $-1/2 \leftrightarrow -3/2$ are allowed, the first occurring at $h\nu = g_{\parallel}\mu_B H_z + 2|D|$, and the latter at $h\nu = g_{\parallel}\mu_B H_z - 2|D|$. Three lines are observed, then, when the magnetic field is parallel to the axis of quantization: a central line, flanked by lines at $\pm 2|D|$.

Thus, zero-field splittings which are of the size of the microwave quantum may be determined with ease and high precision, often even at room temperature, when the conditions described above are realized. As has been implied, however, this simple experiment does not determine the sign of D, but only its magnitude. The sign of D has occasionally been obtained by measuring the spectrum over a wide temperature interval.

It frequently happens that g is anisotropic, and the values of $g_{\parallel}$ and $g_{\perp}$ must be evaluated independently. In rhombic situations, there may be three g-values, g_x, g_y, and g_z.

Now, if there are two resonance fields given by

$$h\nu = g_{\parallel}\mu_B H_{\parallel} \qquad \text{and} \qquad h\nu = g_{\perp}\mu_B H_{\perp},$$

and if $g_{\parallel}$ is measurably different from $g_{\perp}$, then $H_{\parallel}$ will differ from $H_{\perp}$ since $h\nu$ is assumed to be held constant by the experimental equipment. For the sake of argument, let $g_{\parallel} = 2$ and $g_{\perp} = 1$, corresponding, in a particular spectrometer, to typical values of resonance fields of $H_{\parallel} = 3400$ gauss and $H_{\perp} = 6800$ gauss. Define θ as the angle between the principal (parallel) axis of the sample and the external field, and so the above results correspond, respectively, to $\theta = 0^\circ$ and 90°. At intermediate angles, it can be shown that

$$g_{eff}^2 = g_{\parallel}^2 \cos^2\theta + g_{\perp}^2 \sin^2\theta, \tag{3.24}$$

where g_{eff} is merely the effective or measured g-value. A typical set of data for this system would then appear as in Figure 3.16, where both g_{eff}^2 and the resonant field are plotted for the above example as functions of θ. A fit of the experimental data over a variety of angles leads then to the values of both $g_{\parallel}$ and $g_{\perp}$.

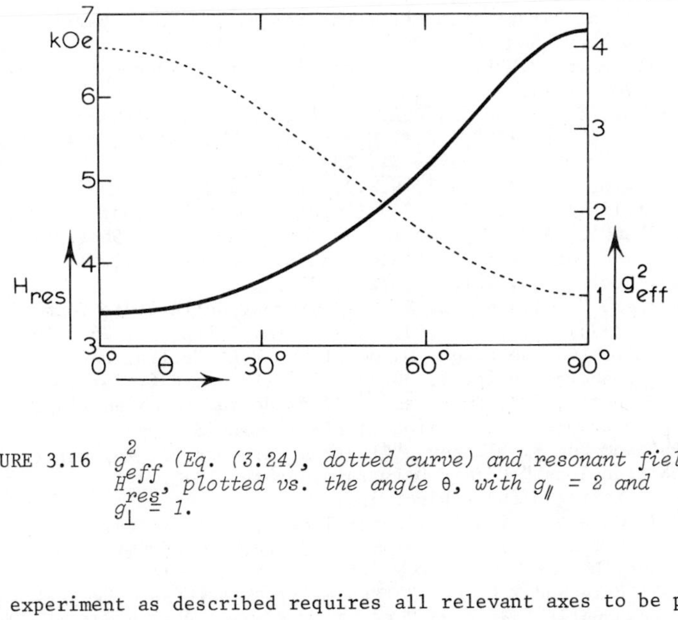

FIGURE 3.16 *g_{eff}^2 (Eq. (3.24), dotted curve) and resonant field, H_{res}, plotted vs. the angle θ, with $g_{\parallel} = 2$ and $g_{\perp} = 1$.*

The experiment as described requires all relevant axes to be parallel, which in turn requires a high-symmetry crystal. Commonly, systems of interest are of lower symmetry, and lines will overlap or several spectra will be observed simultaneously. Methods of solving problems such as these are described elsewhere (17).

REFERENCES

1. H.H. Wickman, M.P. Klein, and D.A. Shirley, J. Chem. Phys. 42, 2113 (1965).
2. C.J. Ballhausen, "Introduction to Ligand Field Theory," McGraw-Hill, New York, 1962.
3. J.W. Stout and W.B. Hadley, J. Chem. Phys. 40, 55 (1964). See also R.A. Fisher, E.W. Hornung, G.E. Brodale, and W.F. Giauque,J. Chem. Phys. 46, 4945 (1967).
4. J.N. McElearney, R.W. Schwartz, A.E. Siegel and R.L. Carlin, J. Am. Chem. Soc. 93, 4337 (1971).
5. M.W. Zemansky, "Heat and Thermodynamics," McGraw-Hill, New York, Ed. 5, 1968.
6. J.H. Van Vleck, "The Theory of Electric and Magnetic Susceptibilities," Oxford University Press, Oxford, 1932.
7. M. Kotani, J. Phys. Soc. Japan 4, 293 (1949).
8. J.N. McElearney, R.W. Schwartz, S. Merchant and R.L. Carlin, J. Chem. Phys. 55, 466 (1971).
9. S. Mitra, "Transition Metal Chemistry," Ed. R.L. Carlin, Marcel Dekker,New York, Vol. 7, 1972, p. 183.
10. R.L. Carlin, "Transition Metal Chemistry," Ed. R.L. Carlin, Marcel Dekker, New York, Vol. 1, 1965, p. 1.
11. A. Abragam and M.H.L. Pryce, Proc. Roy. Soc. (London) A206, 173 (1951).
12. K.W. Mess, E. Lagendijk, D.A. Curtis, and W.J. Huiskamp, Physica 34, 126 (1967); R.F. Wielinga, H.W.J. Blöte, J.A. Roest and W.J. Huiskamp, Physica 34, 223 (1967).
13. H.A. Algra, L.J. de Jongh, H.W.J. Blöte, W.J. Huiskamp, and R.L. Carlin, Physica 82B, 239 (1976); J.N. McElearney, S. Merchant, G.E. Shankle, and R.L. Carlin, J. Chem. Phys. 66, 450 (1977).
14. G.W. Inman, Jr., W.E. Hatfield, and E.R. Jones, Jr., Inorg. Nucl. Chem. Lett. 7, 721 (1971).
15. F. Varret, Y. Allain, and A. Miedan-Gros, Solid State Comm. 14, 17 (1974); F. Varret, J. Phys. Chem. Solids 37, 257 (1976).
16. J.J. Smit, L.J. de Jongh, D. de Klerk, R.L. Carlin, and C.J. O'Connor, Physica B, 86-88, 1147 (1977); R.L. Carlin, C.J. O'Connor, J.J. Smit, and L.J. de Jongh, to be published.
17. The most complete discussion of EPR is provided by: A. Abragam and B. Bleaney, "Electron Paramagnetic Resonance of Transition Ions," Oxford University Press, 1970. Popular texts on the subject are:
    a) B.R. McGarvey, Transition Metal Chem. 3, 90 (1966).
    b) J.W. Orton, "Electron Paramagnetic Resonance," Iliffe Books, London, 1968.
    c) G.E. Pake and T.L. Estle, "Physical Principles of Electron Paramagnetic Resonance," Ed. 2, W.A. Benjamin, Reading, Mass., 1973.
    d) A. Carrington and A.D. McLachlan, "Introduction to Magnetic Resonance," Harper and Row, 1967.

CHAPTER IV

DIMERS AND CLUSTERS

A. INTRODUCTION

Probably the most interesting aspect of magnetochemistry concerns the interactions of magnetic ions. The remainder of this book is largely devoted to this subject, beginning in this chapter with the simplest example, that of a dimer. The principles concerning short range order that evolve here are surprisingly useful for studies on more extended systems.

The previous discussion has considered primarily the Hamiltonian

$$\mathcal{H} = g\mu_B\vec{H}\cdot\vec{S} + D[S_z^{\,2} - 1/3\ S(S+1)] \tag{4.1}$$

which describes only single ion effects. In other words, the properties of a mole of ions have followed directly from the energy levels of the constituent ions, the only complication arising from thermal averaging. Now, let two ions interact through an intervening ligand atom as illustrated in Figure 4.1; this so-called superexchange mechanism, linear in this example, is generally accepted as the most important source of metal-metal interactions or magnetic exchange in insulating compounds of the transition metal ions. Following Martin (1), we consider a three-orbital, four-electron problem : a d_{z^2} electron on each metal ion, and a pair of $2p_z$ electrons on the diamagnetic anion which lies on the metal-metal axis. The ground orbital state is a spin-singlet if the metal ion spins are antiparallel (a) or a spin triplet if parallel (b). The simplest modification of this situation arises from the mixing of small amounts of excited states into the ground states (a) and (b). In particular, the transfer of one electron from a ligand $2p_z$ orbital into the half-filled

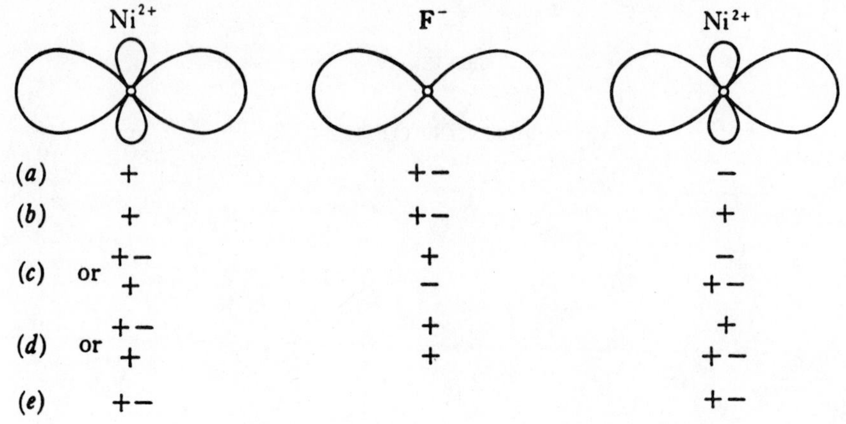

FIGURE 4.1 *Typical orbitals participating in a linear (180°) superexchange pathway between two Ni^{2+} ions (d_{z^2}) via a F^- ion ($2p_z$) as in $KNiF_3$. From Ref. 1.*

d_{z^2} orbital on one metal ion will yield excited orbital singlets as in (c) or triplets as in (d). This amounts to a configurational mixing of states such as M^+ O M^{2+}. The lone electron residing on O in (c) and (d) can now exchange-couple with the lone $3d_{z^2}$ electron on an adjoining M^{2+}. The preferred sign for this coupling is antiferromagnetic (the states are not orthogonal) so that the singlet states (c) are lower in energy than the triplet states (d). The resulting configurational mixing with the ground state leads to stabilization of the (antiferromagnetic) singlet state (a) with respect to the (ferromagnetic) triplet (b).

This model, which has been applied to a wide variety of systems (1,2) is somewhat similar to that which is used for explaining spin-spin coupling in NMR spectroscopy (3). The usual Hamiltonian in use for the metal-metal exchange interaction in magnetic insulators is of the form

$$\mathcal{H}' = -2 \sum_{\substack{i,j \\ i \neq j}} J_{ij} \vec{S}_i \cdot \vec{S}_j \qquad (4.2)$$

where the sum is taken over all pair-wise interactions of spins i and j in a lattice. For the moment, we shall restrict our attention to dimers, and thereby limit the summation to the two atoms 1 and 2 in the dimeric molecule, so that

$$\mathcal{H}' = -2J \vec{S}_1 \cdot \vec{S}_2 \qquad (4.3)$$

This is called an isotropic or Heisenberg Hamiltonian, and we adopt the convention that negative J refers to antiferromagnetic (spin-paired or singlet, in a dimer) interactions, and positive J refers to ferromagnetic (spin-triplet) interactions. The exchange constant J takes energy units and measures the strength of the interaction. The reader must be careful in comparing different authors, for a variety of conventions (involving the negative sign and even the factor of 2) are in use.

B. ENERGY LEVELS AND SPECIFIC HEATS

An antiferromagnetic interaction of the type given in Eq. (4.3), when applied to two ions each of spin-1/2 gives a spin-singlet ground state and a spin triplet 2J in energy above the singlet (Fig. 4.2). Naturally, if the interaction were ferromagnetic, the diagram is simply inverted. For an external field applied along the z-axis of the pair, the complete Hamiltonian will be taken as

$$\mathcal{H} = g\mu_B S_z' H_z - 2J\, \vec{S}_1 \cdot \vec{S}_2 \tag{4.4}$$

where S_z' is the operator for the z-component of <u>total</u> spin of the pair. The eigenvalues of $\mathcal{H}$ are

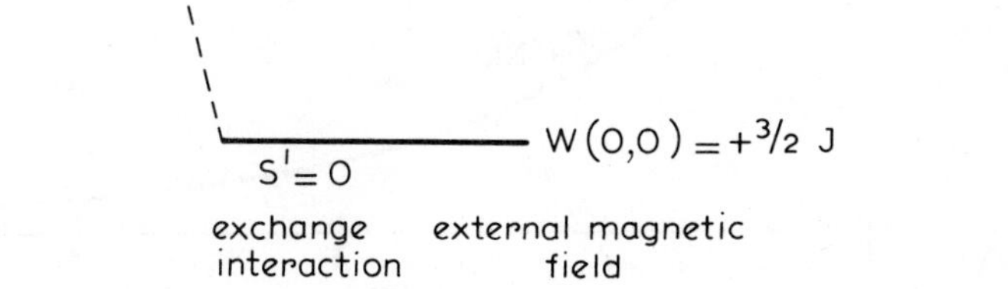

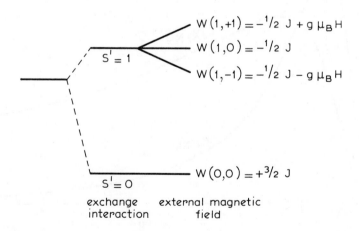

FIGURE 4.2 *Energy-levels for a pair of spin-1/2 ions undergoing magnetic exchange.*

$$W(S',m_S') = g\mu_B m_S' H_z - J[S'(S'+1) - 2S(S+1)]$$

For the dimer, $S = 1/2$, $S' = 0$ or 1, and $m_S' = S', S' - 1, -S'$. The energy level diagram is as shown in Figure 4.2. It is assumed for the moment that there is no anisotropy of exchange interaction.

In the limit of zero magnetic field, the partition function is simply

$$Z = 1 + 3\exp(2J/kT)$$

and the derived heat capacity is quickly calculated as

$$C = \frac{12R(J/kT)^2 \exp(2J/kT)}{[1 + 3\exp(2J/kT)]^2} \tag{4.5}$$

which is of course of exactly the same form as a Schottky specific heat. The reader must be careful in counting the sample; Eq.(4.5) as written refers to a mole of ions, not a mole of dimers. In Figure 4.3, taken from

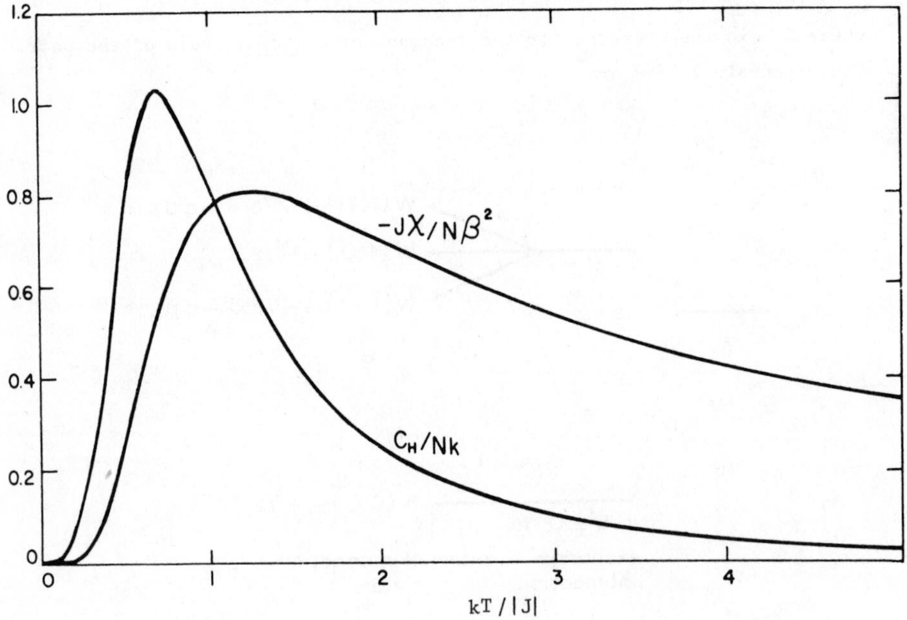

FIGURE 4.3 *Magnetic susceptibility and specific heat of a dimer as a function of kT/|J| for S = 1/2. From Ref. 4.*

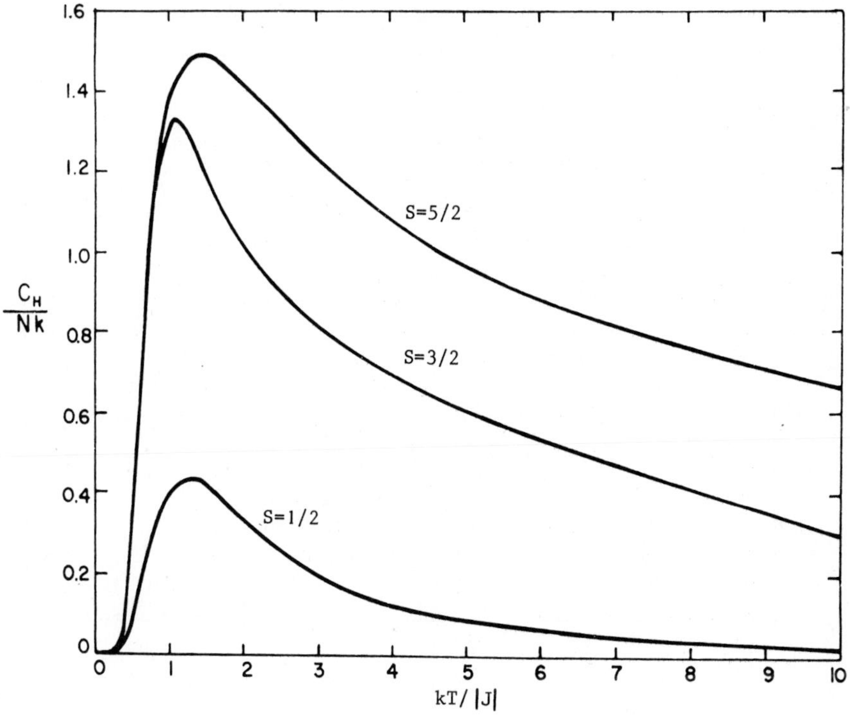

FIGURE 4.4 *Specific heat vs. kT/|J| for antiferromagnetically coupled (J < 0) trimers of S = 1/2, 3/2, 5/2. From Ref. 4.*

Smart (4), we illustrate the general behavior of this curve, and in Figure 4.4, also taken from Smart, the behavior of antiferromagnetically coupled trimers of, respectively, spin 1/2, 3/2 and 5/2 are compared. Note that the position of the maxima and the low temperature behavior in the latter figure are approximately independent of S.

There seems to be but one example available of a compound that has a specific heat which follows Eq.(4.5), and that is $Cu(NO_3)_2 \cdot 2\frac{1}{2}H_2O$ (5). For most compounds containing exchange-coupled dimers, 2J is so large that the overlap of the magnetic contribution with the lattice contribution is so serious as to prevent their separation and identification. In the case of $Cu(NO_3)_2 \cdot 2\frac{1}{2}H_2O$, however, the singlet-triplet separation is only about 5.2 K (5), with the singlet lower, and the broad maximum (Fig. 4.5) is easily discernible.

The investigation of this system was even more interesting because the effect of a magnetic field on the specific heat was also examined. In

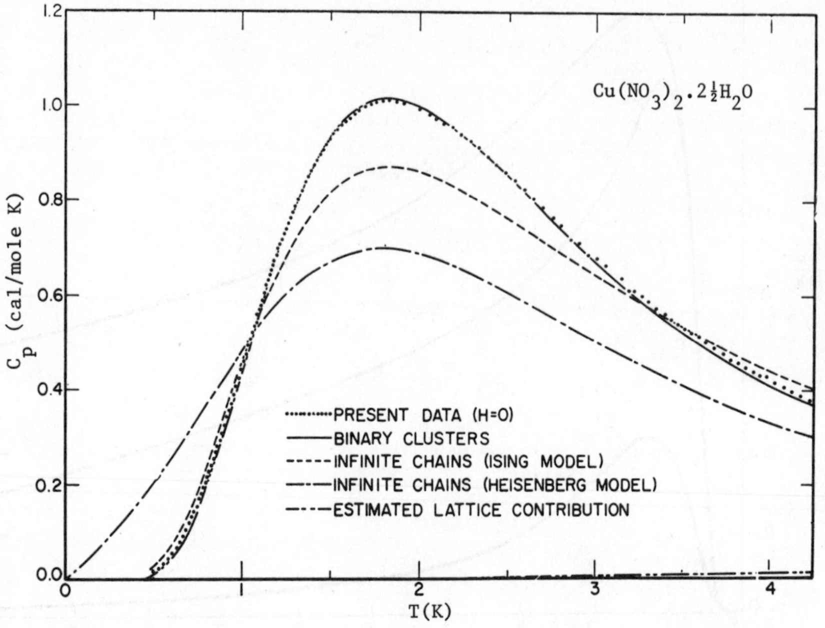

FIGURE 4.5 *Heat capacity of polycrystalline $Cu(NO_3)_2 \cdot 2\frac{1}{2}H_2O$ in zero external magnetic field. From Ref. 5.*

this case, the far right side of Figure 4.2 is applicable, and the partition function becomes (in the case of isotropic exchange),

$$Z = 1 + \exp[(2J-g\mu_B H)/kT] + \exp(2J/kT) + \exp[(2J+g\mu_B H)/kT]$$

$$= 1 + \exp(2J/kT)\,[1 + 2\cosh(g\mu_B H/kT)] \qquad (4.6)$$

Inserting the best-fit parameters of $g = 2.13$ and $2J/k = -5.18$ K, a fit to this model at H = 8.7 kGauss is illustrated in Figure 4.6. The agreement is striking. None of the other models applied to the data, such as infinite chains of atoms, fit either the zero-field or applied-field specific heat results. It is therefore all the more remarkable to find (6) that $Cu(NO_3)_2 \cdot 2\frac{1}{2}H_2O$ does not contain binary clusters of metal atoms, but is actually chain-like with bridging nitrate groups, Figure 4.7. The chain is not linear but crooked and this may be why the short-range or dimer ordering occurs here. The magnetic susceptibility of a dimer, which is discussed in the next section, has a broad maximum at temperatures compa-

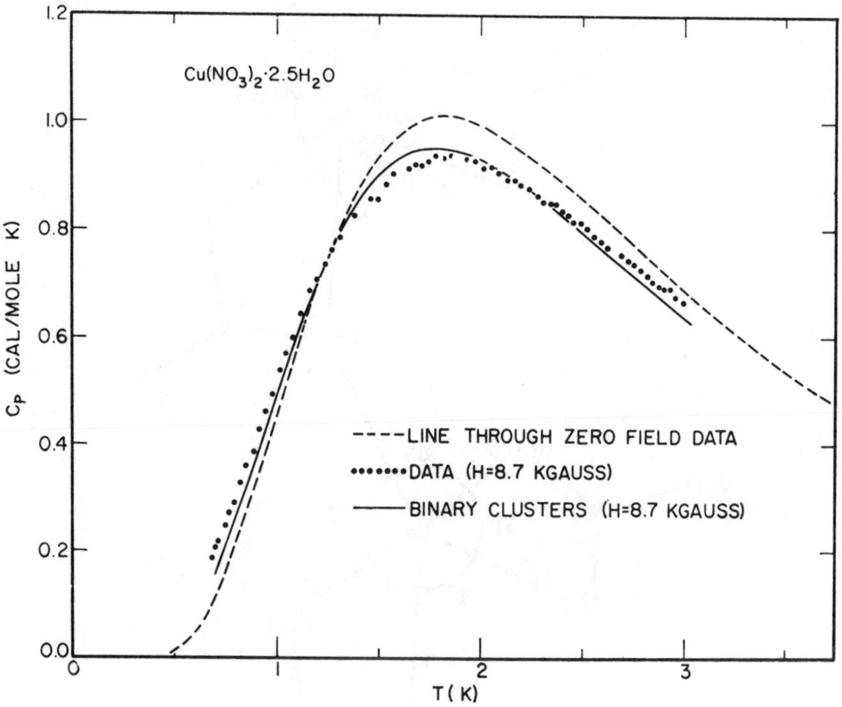

FIGURE 4.6 *Heat capacity of polycrystalline* $Cu(NO_3)_2 \cdot 2\frac{1}{2}H_2O$ *in an applied field of 8.7 kG. From Ref. 5.*

rable to the singlet-triplet separation. The susceptibilities of $Cu(NO_3)_2 \cdot 2\frac{1}{2}H_2O$ nicely fit the theory of the next section with the same exchange constant as derived from the specific heat measurements (7).

C. MAGNETIC SUSCEPTIBILITIES

Begin first with the application of Van Vleck's equation to the energy level situation sketched in Figure 4.2. The isothermal magnetic susceptibility per mole of dimers is readily calculated as:

$$\chi = \frac{2N(g\mu_B)^2 \exp(J/2kT)}{3\exp(J/2kT) + \exp(-3J/2kT)} = \frac{2Ng^2\mu_B^2/kT}{3 + \exp(-2J/kT)}$$

or

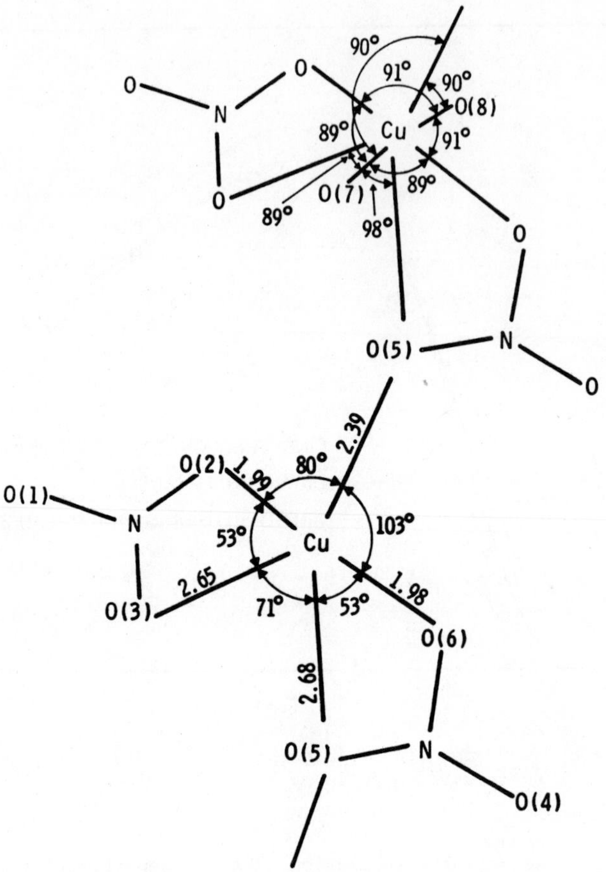

FIGURE 4.7 *A schematic drawing of the crooked chains which link the copper ions in* $Cu(NO_3)_2 \cdot 2\frac{1}{2}H_2O$. *The two water oxygen atoms above the lower copper atom have not been included. From Ref. 6.*

$$\chi = (2Ng^2\mu_B^2/3kT)\,[1 + \tfrac{1}{3}\exp(-2J/kT)\,]^{-1} \qquad (4.7)$$

Since $N\mu_B^2/3k \approx 1/8$, and setting $g = 2$ as a first approximation,

$$\chi = (1/T)\,[1 + \tfrac{1}{3}\exp(-2J/kT)\,]^{-1}$$

Note that, for negative J, $\chi \to 0$ both for $T \to 0$ and $T \to \infty$; clearly, χ must have a maximum. The easiest way to find the temperature, T_m, of

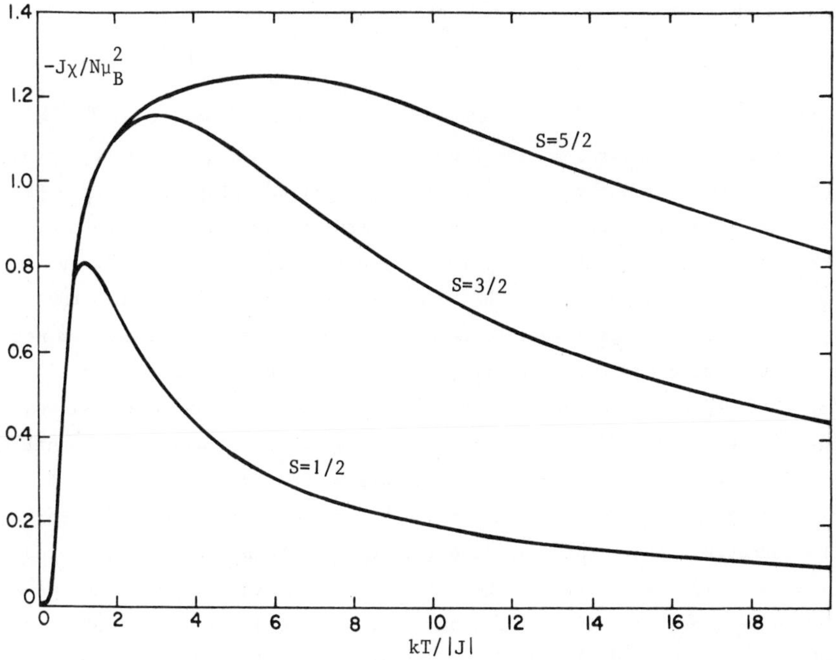

FIGURE 4.8 *Magnetic susceptibility of (antiferromagnetic) pairs vs. kT/|J| for S = 1/2, 3/2, 5/2. From Ref. 4.*

this maximum is to set $\partial \ln\chi/\partial T = 0$. One finds, with the definitions used here, that the maximum occurs at $J/kT_m \approx -4/5$. For $J << kT$ (or $T >> T_m$), the susceptibility follows a Curie-Weiss law, $\chi = (3/4)/(T-\theta)$ with $\theta = J/2k$, which illustrates the well-known connection between a Curie-Weiss θ and the exchange interaction. It has already been pointed out that this is not a unique relationship, however. Note also that $\chi \to 0$ as $-J \to \infty$ or, as we would expect, as the energy state in which the dimer is paramagnetic gets further away (higher in energy) because of a stronger exchange coupling, the susceptibility at a given temperature must decrease. On the other hand, if J should be positive, an S = 1 spin-only Curie law susceptibility is obtained as 2J/kT becomes large.

The reduced susceptibility for a pair of ions was illustrated in Figure 4.3, and a similar calculation for S = 1/2, 3/2 or 5/2 pairs is illustrated in Figure 4.8. While the low temperature behavior in the three cases is the same, note that the temperature of maximum χ increases with S.

Examples to which Eq. (4.7) has been applied are legion (1,4,8), the most famous example being copper acetate, $Cu(OAc)_2 \cdot H_2O$.

D. COPPER ACETATE AND RELATED COMPOUNDS

Bleaney and Bowers (9) first brought copper acetate to the attention of both chemists and physicists in 1952 by their investigation of the EPR spectrum of the pure crystal. They were drawn to the compound because, although most copper salts had relatively straightforward magnetic behavior for S = 1/2 systems, it had been reported by Guha (10) that the susceptibility of copper acetate monohydrate passed through a maximum near room temperature and then decreased so rapidly as the temperature fell that it would apparently become zero at about 50 K. No sharp transition had been reported, and the behavior was unlike that found with the usual antiferromagnets. The EPR spectrum was also inconsistent with that of a normal copper salt: at 90 K, a line at X-band (0.3 cm^{-1}) was observed in zero magnetic field. This is inconsistent with the behavior anticipated for an S = 1/2 Kramer's ion. The spectrum was similar but more intense at room temperature, but disappeared at 20 K. The spectra bore certain resemblances to those of nickel(II), which is a spin-1 ion with concomitant zero-field splittings.

The simplest explanation for these results, and the one put forward by Bleaney and Bowers, was that the copper ions must interact antiferromagnetically in pairs. As described in earlier sections of this chapter such a pair-wise exchange interaction yields a singlet ground state and a triplet excited state. In contrast to the example of $Cu(NO_3)_2 \cdot 2\frac{1}{2}H_2O$, the splitting in copper acetate must be very large in order to cause the reported magnetic behavior, and in fact the singlet-triplet separation should be comparable to thermal energy (kT) at room temperature. Zero susceptibility and the lack of an EPR spectrum at low temperatures, when the triplet state has lost its population, follow immediately. This remarkable hypothesis has since been found to be true by a variety of experiments.

The crystal structure of $Cu(OAc)_2 \cdot H_2O$ was reported shortly thereafter by van Niekerk and Schoening (11). Isolated dimers were indeed found to be present. The structure, illustrated in Figure 4.9, has also been refined more recently (12,13). A pair of copper atoms is supported by four bridging acetate groups in a D_{4h} array with two water molecules completing the coordination along the Cu-Cu axis. A similar structure is found for

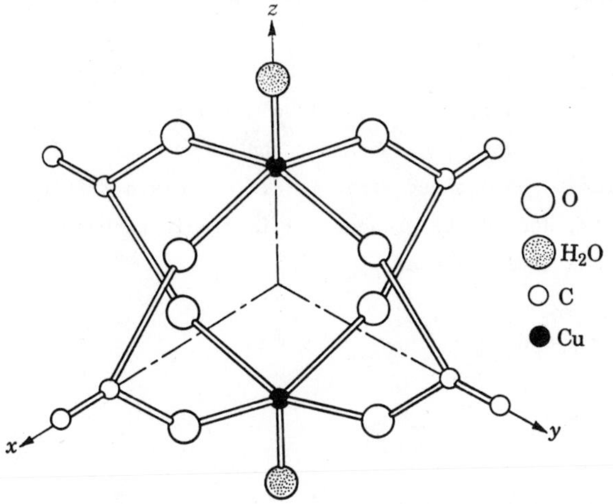

FIGURE 4.9 *Molecular structure of hydrated copper acetate. From Ref. 1.*

the chromous compound (14) but there the exchange interaction is so large that the compound is diamagnetic even at room temperature.

The first EPR results (9) were interpreted in terms of the spin-Hamiltonian

$$\mathcal{H} = g\mu_B\vec{H}\cdot\vec{S} + D[S_z^{\,2} - \tfrac{1}{3}S(S+1)] + E(S_x^{\,2}-S_y^{\,2}) \qquad (4.8)$$

with $D = 0.34 \pm 0.03$ cm^{-1}, $E = 0.01 \pm 0.005$ cm^{-1} and $g_z = g_{\parallel} = 2.42 + 0.03$, $g_x = g_y = g_{\perp} = 2.08 \pm 0.03$. These results apply of course to the excited $S = 1$ state; the Cu-Cu axis serves as the z-axis. The singlet-triplet separation, or isotropic exchange interaction, was estimated, to 20%, as 370 K. More recent results, for both $Cu_2(OAc)_4\cdot 2H_2O$ and the zinc-doped monomers (15) $ZnCu(OAc)_4\cdot 2H_2O$ (the $Cu_2(OAc)_4\cdot 2H_2O$ structure will allow replacement of about 0.5% of Cu by Zn) are $g_z = 2.344 \pm 0.005$, $g_x = 2.052 \pm 0.007$ and $g_y = 2.082 \pm 0.007$. The similarity of these values to those found in the more normal or monomeric copper salts supports the suggestion that the copper atoms are, aside from the exchange interaction, subjected to a normal type of crystalline field.

In the last chapter, the values of the spin-Hamiltonian parameters D and E were correlated with zero-field splittings which were caused by axial crystalline fields. As has been shown by Abragam and Bleaney (16), these parameters may be correlated in the present case with anisotropic exchange.

There are two energy levels considered for the dimer, one with S = 1 which is -2J in energy above the level with S = 0. As has been pointed out above, the lower level does not and cannot make any contribution to an EPR spectrum. The excited state is effectively a triplet, as with other S = 1 states, and in the presence of isotropic exchange and an external field H_z, the levels are at

$$-J/2 + g_z \mu_B H_z,$$
$$-J/2,$$
$$-J/2 - g_z \mu_B H_z.$$

With the usual EPR selection rule of $\Delta m = \pm 1$, two transitions occur, both at

$$h\nu = g_z \mu_B H_z$$

This is independent of the value of J, showing that isotropic exchange has no other effect on the spectrum except at temperatures where $kT/J \approx 1$, when the intensity will no longer vary inversely as the absolute temperature because of the triplet-singlet splitting. The exchange constant, J, may be written

$$J = (1/3)\ [(J_x - J_x') + (J_y - J_y') + (J_z - J_z')] \tag{4.9}$$

where primed components denote the anisotropy in the exchange constant; if the exchange were isotropic,

$$J_x' = J_y' = J_z' = 0, \text{ and } J = (1/3)(J_x + J_y + J_z).$$

Furthermore, a constraint on the anisotropic portion is that

$$J_x' + J_y' + J_z' = 0.$$

The zero-field splitting of the S = 1 state is then due to the anisotropic contribution, and in fact Abragam and Bleaney show that we may associate the parameters of Eqs. (4.8) and (4.9) as

$$D = 3J_z'/4 = 0.34 \text{ cm}^{-1}$$

and

$$E = 1/4(J_x' - J_y') = 0.01 \text{ cm}^{-1}$$

and, recalling that the isotropic term was estimated to be of the order of 370 K (260 cm^{-1}), we see that, relatively, the anisotropic exchange is quite small.

The first careful measurement of the susceptibility of $Cu(OAc)_2 \cdot H_2O$ was carried out by Figgis and Martin (17) and the most recent study (18) is that also of Martin and co-workers. The results are in complete accord with the above discussion and may be fitted by Eq. (4.7), as illustrated in Figure 4.10, for a more accurate evaluation of J. It was found (17) that T_{max} = 255 K, and that -2J/k = 480 K when the Hamiltonian of Eq. (4.3) is used. What was even more fascinating, Figure 4.11, is that anhydrous copper acetate behaves quite similarly, with T_m = 270 K and -2J/k = 432 K. The conclusion that anhydrous copper acetate not only retains the gross molecular structure of the hydrate but also behaves antiferromagnetically in the same fashion is inescapable. Furthermore, replacement of the axial water molecule by, for example, pyridine results in a similar magnetic (19) and crystallographic (20) situation. In this case, the exchange interaction becomes -2J/k = 481 K, and similar results are obtained with substituted pyridines as well as when acetate is replaced by such alkanoates as propionate and butyrate. The hydrated and anhydrous formates, however, are distinctly different, both in structure and magnetic properties (21).

The great success of this magnetic model for copper acetate and its analogues has led to its application to almost any copper compound that has a subnormal magnetic moment at room temperature. In many cases, the structure has been correctly deduced (22). But, as was the case with hydrated copper nitrate, magnetic properties turn out to be a more fallible indicator of crystal structure than does X-ray crystallography (8). For example, the aniline adducts of copper acetate behave similarly to the other dimers described above (19), though the singlet-triplet energy was reduced substantially so that the temperature of maximum susceptibility was not measured directly. The parameter -2J/k was estimated as

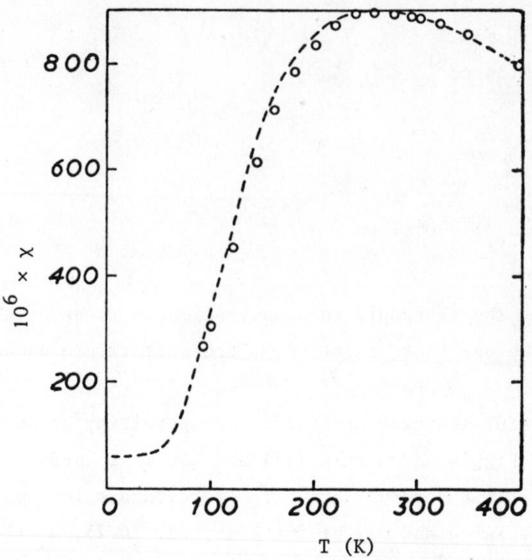

FIGURE 4.10 *Experimental and calculated magnetic susceptibilities of $Cu(OAc)_2.H_2O$. From Ref. 17.*

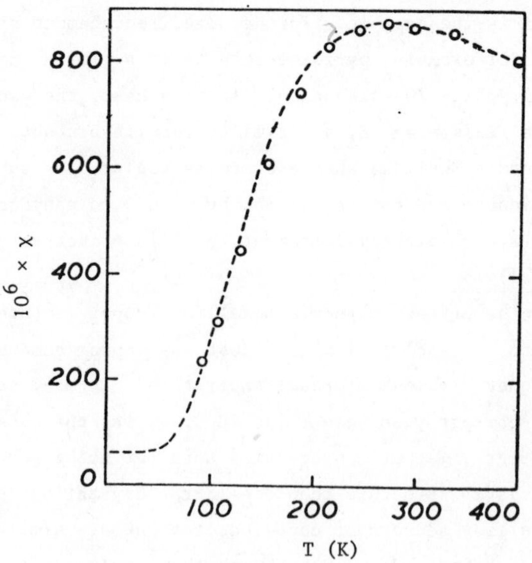

FIGURE 4.11 *Experimental and calculated magnetic susceptibilities of anhydrous copper acetate. From Ref. 17.*

only about 150 K for both the m- and p- toluidine adducts of copper butyrate. It has since been shown (23) that the p-toluidine adduct of copper acetate is not binuclear, but instead polymeric chains are formed. The powder susceptibility of $Cu(OAc)_2 \cdot 2$p-toluidine$\cdot 3H_2O$ was fitted (23) as an Ising chain (Chapt. VI) rather then to the Bleaney and Bowers relationship, but, as has been pointed out (24), copper is usually a Heisenberg or isotropic ion. A more complete set of measurements is required before one can be satisfied that this system is well-understood. Furthermore (25), the p-toluidine adduct of copper(II) propionate, which is composed of one-dimensional polymeric chains, exhibits a magnetic susceptibility (over a limited range of temperatures) which fits the Bleaney-Bowers scheme quite well.

A further question remains to be discussed, and that is what is the mechanism of spin-spin coupling in the copper acetate compounds? A number of models have been presented (1), and there are probably more theoretical papers on the subject than experimental contributions. Just the same, there is no model acceptable to everyone interested in the problem, and there is probably no experimental test available yet that can settle the disagreements.

The first bonding scheme, presented in the first (17) magnetic study, proposed that there was a direct bond formed between the two metal atoms. After all, the copper atoms are only 2.616 Å apart in $Cu(OAc)_2 \cdot H_2O$. It was proposed that a δ-bond was formed by a lateral overlap of two $d_{x^2-y^2}$ orbitals, and a variety of spectral data remain consistent with this proposal (1). Superexchange interaction through the carboxylate groups has also been widely suggested as the more important contributor. A great deal of indirect information tends to favor this scheme. For example, though the structure of copper propionate p-toluidine is not the usual binuclear one, the susceptibility behaves similar to that of the binuclear molecules, and yet the Cu-Cu separation is increased to a range of 3.197 to 3.341 Å. This distance, and the arrangement of the chains of atoms, make a direct overlap seem unlikely. Furthermore, the quinoline adduct of copper(II) trifluoroacetate is binuclear with the molecular structure common to the copper acetate dimers (26). The Cu-Cu distance is 2.886 Å, a full 0.272 Å longer than the corresponding distance in $Cu(OAc)_2 \cdot H_2O$. The susceptibility obeys the Bleaney-Bowers relationship over the range 80-300 K with $g = 2.27$ and $2J = -310\ cm^{-1}$ ($2J/k = -446$ K). The large difference in Cu-Cu separation between the magnetically similar acetate and trifluoroacetate adduct demonstrates that the metal-metal distance in these dimers is not an important factor in determining the strength of the Cu-Cu interaction, which in turn lends further weight to the importance of the superexchange mechanism.

E. SOME OTHER DIMERS

The bis(cyclopentadienyl)titanium(III) halides form another series (27) of binuclear molecules of the type $[(\eta\text{-Cp})_2\text{Ti-X}_2\text{-Ti-}(\eta\text{-Cp})_2]$, with two halide bridging atoms. The series is interesting because X may be any one of the four halide ions. The magnetic susceptibilities indeed follow Eq. (4.7), the singlet-triplet equation, with -2J values of the order of 62 cm^{-1} (F), 159-186 cm^{-1} (Cl), 276 cm^{-1} (Br) and 168-179 cm^{-1} (I). The relative behavior of the chlorine and bromine derivatives follows a common, but not universal, trend, but in the absence of detailed structural data (especially concerning the dimensions and angles of the Ti_2X_2 core), the origin of the trends in the Ti-Ti interaction must remain uncertain.

An unusual example of 1,3-magnetic exchange has recently been reported (28). The compounds are of the type $[Cp_2Ti]ZnCl_4$, which contain linear units of the sort

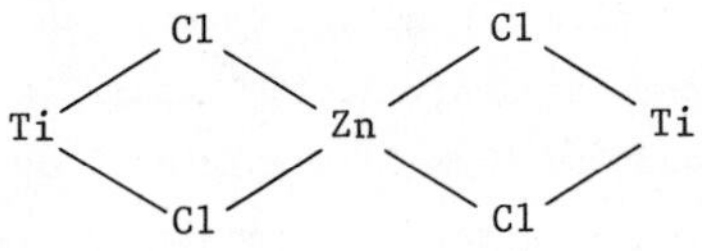

with a Ti-Zn distance of 3.420 Å, a Zn-Cl-Ti angle of 89.9°, a Cl-Ti-Cl angle of 82.1° and a Ti-Zn-Ti angle of 173.4°. The Ti-Ti distance is 6.828 Å. Similar compounds with Cl replaced by Br, and Zn replaced even by Be were also reported. Despite the intervening diamagnetic MX_4 unit, the compounds follow the Bleaney-Bowers relation with 2J/k of the order -20 to -40 K. An examination of the crystal packing suggested to the authors that intermolecular exchange was not the source of the antiferromagnetic behavior.

A variety of other dimers have been examined and reviewed (1). In Figure 4.12, we reproduce (1) the reduced reciprocal magnetic susceptibility calculated as a function of reduced temperature for a series of dimeric metal complexes as a function of both spin and the sign of the exchange interaction. Broad maxima which shift to higher temperature as the spin increases, are found when the exchange constant is negative (anti-

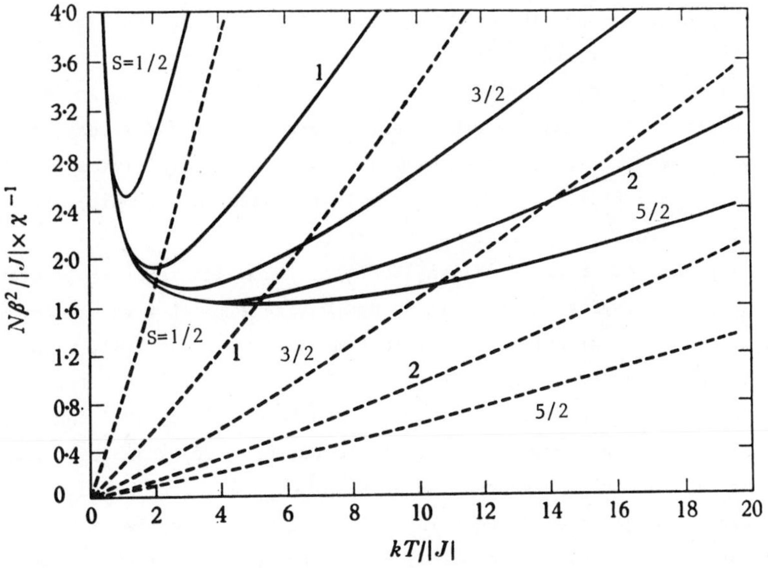

FIGURE 4.12 *Reciprocal magnetic susceptibility vs. kT/|J| for S = 1/2, 3/2, 2, 5/2. Full curves, J negative; broken curves, J positive. From Ref. 1.*

ferromagnetic). As will be discussed in the next chapter, these maxima do not indicate long range or antiferromagnetic ordering, and it is therefore wrong to call T_{max} a Néel or critical temperature. Considerable confusion exists in the literature on this point. Maxima are not obtained when the exchange interaction is positive, and therefore a very careful fitting of theory to experiment is required in order to prove that positive exchange coupling is in fact occurring. Broad maxima in the specific heats are required in both situations (4,8).

We will restrict further discussion here to one more system (29), $Ni(\underline{en})_2X_2$, which is interesting because the double-halide bridge is characteristic of many chemical and magnetic chain systems (Chapt. VI), and because of the complications raised by introducing a metal ion with spin S = 1. The compounds $[Ni_2en_4Cl_2]Cl_2$ and $[Ni_2en_4Br_2]Br_2$ are di-(μ-halo) bridged dimers in which a ferromagnetic coupling appears to take place via an approximately 90° Ni-X-Ni (X = Cl, Br) interaction. Susceptibilities of powdered samples were measured over a wide temperature range (1.5-300 K) and magnetic field range (1-15.3 kOe) and compared to calculations carried out similarly to those described above. The effects of interdimer interaction were examined in a molecular field approximation (see Chapt. V), and the effects of the likely large zero-field splittings were also examined. Unfortunately, in its effect on the dimer susceptibility, zero-field splitting is qualitatively similar to an antiferromagnetic interdimer interaction. These systems are good examples of the situation where so many factors are at work that single crystal susceptibilities (the calculated susceptibilities for the model are quite anisotropic) are required for a final analysis of the situation.

The remaining dimer is $[Ni_2en_4(SCN)_2]I_2$, which has a di(μ-thiocyanato) structure. The powder susceptibility again fits a model with a molecular ground state of total spin 2 (i.e., ferromagnetic interaction), with an intracluster exchange integral of 8.6 K. This is an important result because the Ni atoms are far apart (5.8 Å) and because they are connected by two rather long three-atom bridges. The occurrence of exchange coupling with the same order of magnitude in both $[Ni_2en_4X_2]X_2$ and $[Ni_2en_4(SCN)_2]I_2$, in spite of the great difference in Ni-Ni distance (which is probably about 3.5 Å in the chloride complex), demonstrates the relative unimportance of the metal-metal distance in determining the strength of exchange interactions, so long as there exist appropriate pathways for exchange coupling through bridging ligands.

F. EPR MEASUREMENTS

One of the principal applications of EPR to transition metal ions is to determine spin-Hamiltonian parameters (16). When metal ions are put into diamagnetic hosts in small concentration, this is one of the most accurate ways of determining g-values and zero-field splittings that are relatively small. Exchange and other effects often limit the amount of information that can be obtained by EPR on concentrated materials, and so a recent study (30) of the $Cr_2Cl_9^{3-}$ ion as the Cs^+, Et_4N^+ and Pr_4N^+ salts is of considerable interest and illustrates a procedure of broad application.

The complex ion is formed by the sharing of a face between two adjacent $CrCl_6$-octahedra. Thus, the metal atoms are bridged by three chloride ions, and are 3.12 Å apart in the Cs salt. The early powder susceptibility measurements (31) go down only to 80 K and so those data are not very sensitive to exchange, which was estimated to be $2J/k = -10$ K for the usual Hamiltonian, $\mathcal{H} = -2J\vec{S}_1 \cdot \vec{S}_2$. Under such an interaction which is antiferromagnetic in sign, two chromium(III) ions of spin-3/2 form a manifold of four levels of total spin 0 (at 15/2 J), 1 (at 11/2 J), 2 (at 3/2 J) and 3 (at -9/2 J), where the terms in parentheses refer to the energies of the different levels with respect to the free ion levels. With negative J, the S = 0 level lies lowest, the other levels are expected to have a Boltzmann population, and so a non-Curie susceptibility that goes to zero as $T \to 0$ is predicted. Each of the several levels is expected to give an EPR spectrum characteristic of the total spin of the particular level, as in the case of copper acetate, with a zero-field splitting that is once again due to anisotropy in the exchange and dipole-dipole forces. For the Cs^+ salt, only the S = 1 level was observed, while the S = 2 state was the only one observed for $(Et_4N)_3Cr_3Cl_9$. Spectra were observed from all the manifolds of different total-spin with the tetra-n-propylammonium salt, which allowed a large deviation from the Landé interval rule to be observed. In particular, in addition to the usual bilinear term in $\vec{S}_1 \cdot \vec{S}_2$, it was found necessary for the data analysis to add smaller terms in $(\vec{S}_1 \cdot \vec{S}_2)^2$ and even $(\vec{S}_1 \cdot \vec{S}_2)^3$. These latter isotropic terms may arise not so much from superexchange interaction but from the effects of such phenomena as exchange striction, the interaction of exchange with the elastic constants of the crystalline material. The interaction constants were estimated by the temperature variation of the intensity, and 2J/k = -16 to -20 K was found for the three compounds, with very little exchange coupling between the pairs. Recent susceptibility data are consistent with this interpretation (32).

G. CLUSTERS

The methods described above are of course applicable to any discrete cluster of magnetic atoms. Trimers and tetramers are well-known (1,8,33) and we are limited only by the ease of chemical synthesis and the increasing mathematical complexity of the problem as the symmetry of the cluster

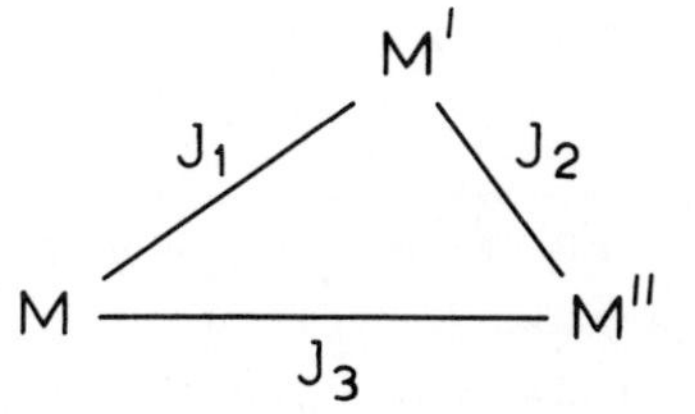

FIGURE 4.13

Three metal atoms with inequivalent magnetic exchange.

decreases and the interactions increase in both number and kind. Thus, consider the metal atom triad of Figure 4.13, with three exchange constants. The metal atoms may or may not be alike, zero-field splittings may contribute if the spin is greater than 1/2, and the exchange constants may or may not be the same. The more parameters that are introduced, the more

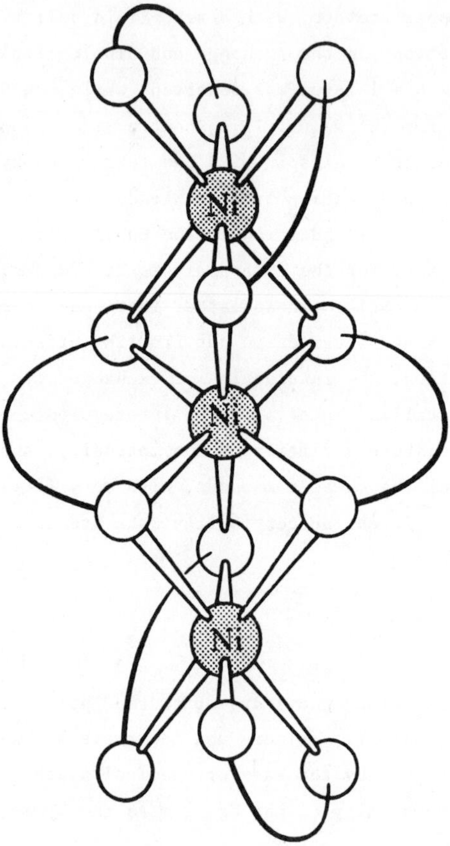

FIGURE 4.14 *Molecular structure of* $[Ni\ acac_2]_3$. *From Ref. 34.*

detailed must be the experimental data in order to separate the different factors contributing to the magnetic properties.

An interesting example is offered by linear trimeric bis(acetylacetonato)nickel(II), $[Ni\ acac_2]_3$, which is illustrated in Figure 4.14.

For the isolated trimer, the exchange Hamiltonian may be written (34) as

$$\mathcal{H} = -2J[\vec{S}_1\cdot\vec{S}_2 + \vec{S}_2\cdot\vec{S}_3] - 2J'(\vec{S}_1\cdot\vec{S}_3) \tag{4.10}$$

where J is the exchange integral between adjacent nickel atoms (1,2 and 2,3), and J' is the exchange term between the two terminal nickel atoms (1,3) within the trimer. The spin of the trimeric molecule as well as the order of the energy level manifold depends on the relative values of J and J', as well as the sign of J. For example, for positive J, the ground state is always paramagnetic, but the spin depends on the value of J'/J. For negative J, the ground state is diamagnetic only in the case of $0.5 < J'/J < 2$; otherwise it is paramagnetic. The powder susceptibility data were best fit by a ferromagnetic J/k = 37 K, and an antiferromagnetic J'/k = -10 K. This suggests that the molecular ground state of the trimer should have a total spin of 3.

An interesting series of clusters contains the trimeric ion, $[M_3O(CH_3CO_2)_6(H_2O)_3]^+$. Compounds where M is Cr or Fe have been studied

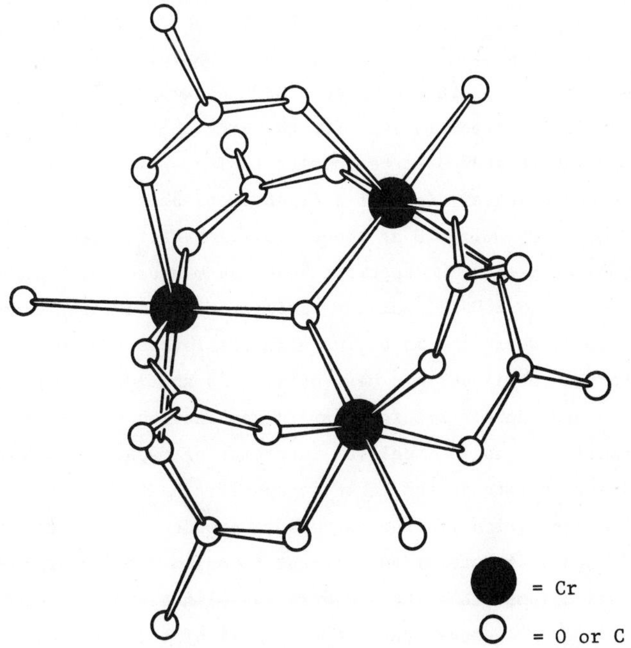

FIGURE 4.15 *Molecular structure of the trimeric cation,* $[Cr_3O(CH_3CO_2)_6(H_2O)_3]^+$. *From Ref. 1.*

most extensively, and the mixed Cr_2Fe compound has also been examined. Some of the problems raised here are quite illuminating, for they illustrate the difficulty involved in studying large clusters.

The molecular structure (35) of $[Cr_3O(CH_3CO_2)_6(H_2O)_3]^+$ is illustrated in Figure 4.15. Each Cr^{3+} ion is nearly octahedrally coordinated by oxygen, and the three metal atoms are arranged, at ambient temperature, in an equilateral triangle about a central oxide anion. Early measurements of the specific heat and susceptibility (36) suggested that an antiferromagnetic interaction occurred among the three ions, and with a Hamiltonian of the form

$$\mathcal{H} = -2J[\vec{S}_1\cdot\vec{S}_2 + \vec{S}_2\cdot\vec{S}_3 + \alpha\vec{S}_1\cdot\vec{S}_3] \qquad (4.11)$$

a fair fit to the data of a number of investigators was obtained with $J/k = -15$ K and α between 1 and 1.25. A value of α different from 1 signifies that the magnetic interactions are more like that of an isosceles triangle than the structural equilaterial triangle. Since the ground state of the system of three spin-3/2 ions corresponds to a total net spin of 1/2, and is therefore paramagnetic, intercluster coupling is expected to occur at sufficiently low temperatures. Susceptibility measurements provide no evidence for this down to 0.38 K, however (1). Attempts to add higher-order interactions (37) to fit the data better to the equilateral triangle model have proved to be invalid (38).

The recent discovery (38) of a crystallographic phase transition for $[Cr_3O(CH_3COO)_6(H_2O)_3]Cl\cdot6H_2O$ at about 210 K serves to make all the previous data analyses suspect. The specific heat was measured over a wide temperature interval (1.5-280 K), but the estimation of the lattice specific heat below 20 K, where the major intracluster interaction occurs, depended on a fit to the total specific heat between 30 and 100 K, which is a questionable procedure. Straightforward application of both the equilateral and isosceles triangle models of intracluster magnetic exchange did not give satisfactory fits to the magnetic specific heat, but the observation of the phase transition led the authors to propose a new model for this system. Though a structural equilateral triangle obtains at room temperature, it was assumed that the symmetry was slightly distorted through the phase transition processes, but the unit cell of the low temperature phase preserves four formula units in two sets of equivalent pairs. In other words, two sets of isosceles triangles were assumed, and an excellent fit to the magnetic specific heat was obtained with the following parameters for the spin-Hamiltonian

$$\mathcal{H} = -2J_o(\vec{S}_1\cdot\vec{S}_2 + \vec{S}_2\cdot\vec{S}_3 + \vec{S}_1\cdot\vec{S}_3) - 2J\vec{S}_2\cdot\vec{S}_3 \tag{4.12}$$

Set 1:	Set 2:
$2J_o/k = -30$ K	$2J_o/k = -30$ K
$2J_1/k = -4.5$ K	$2J_1/k = +1.5$ K

These results are not inconsistent with any other available data, and the presence of two sets of units was in fact confirmed (39) by the analysis of the emission and optical spectra.

The tetranuclear ion, $[Cr_4(OH)_6en_6]^{6+}$ offers another interesting example of the problems associated with the study of clusters. In this case, the major problem was that the early (40,41) magnetic studies were interpreted on the basis of the incorrect structural model. The powder susceptibility of $[Cr_4(OH)_6en_6]I_6\cdot 4H_2O$ over a limited temperature interval is illustrated in Figure 4.16, along with a fit to the data using a model

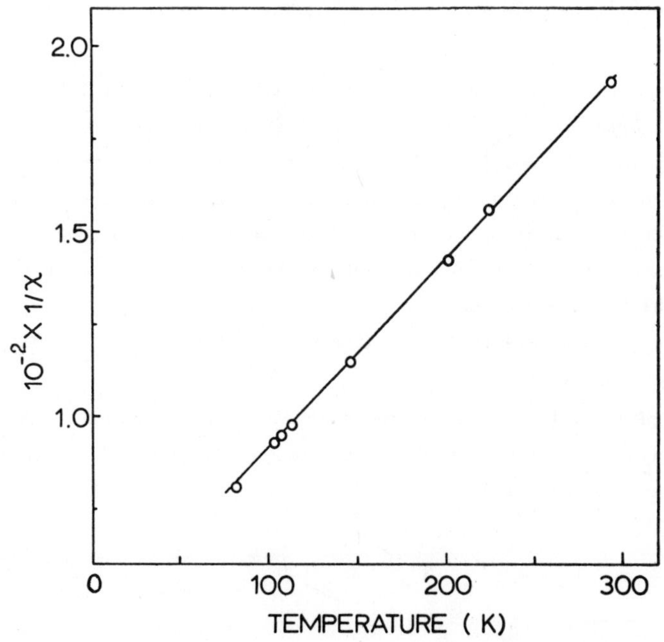

FIGURE 4.16 *Reciprocal susceptibilities of* $[Cr_4(OH)_6en_6]I_6\cdot 4H_2O$. *The line represent calculated values. From Ref. 40.*

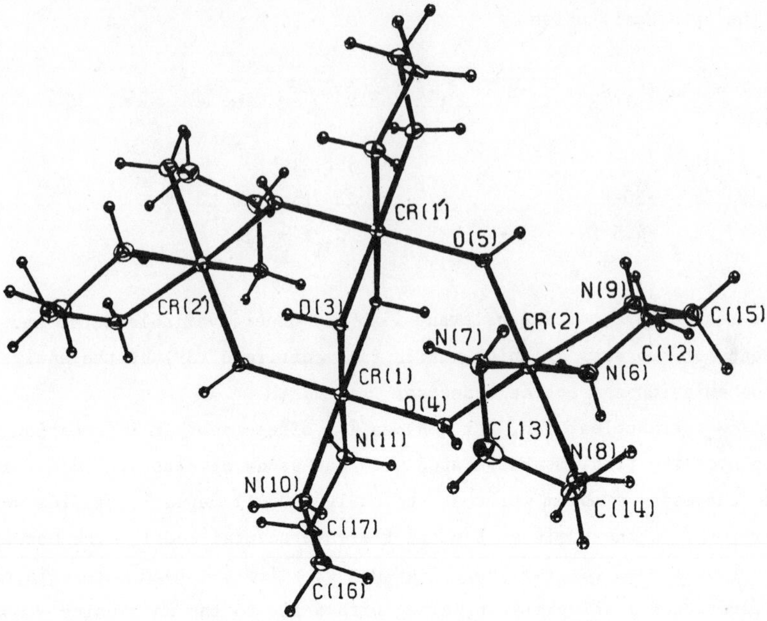

FIGURE 4.17 *A view of the* $[Cr_4(OH)_6en_6]^{6+}$ *cation. From Ref. 43.*

presumed to apply to a trigonal structure with a central Cr(III) atom interacting magnetically with the other three atoms (40); it was later (42) pointed out that the model used was in fact one applicable to a tetrahedral Cr_4 cluster. Data over a wider temperature range were obtained (41) for $[Cr_4(OH)_6en_6](SO_4)_3 \cdot 10H_2O$, a molecule presumed to contain the same tetranuclear cluster, and the authors correctly used the trigonal model to obtain agreement between experiment and theory. However, in this case it was found necessary to add higher-order terms of the form $(\vec{S}_i \cdot \vec{S}_j)^2$ in order to fit the data.

Subsequently, it was discovered that Pfeiffer's cation, $[Cr_4(OH)_6en_6]^{6+}$, as it was found in $[Cr_4(OH)_6en_6](N_3)_6 \cdot 4H_2O$, is in fact a Cr_4 planar rhomboid (43),

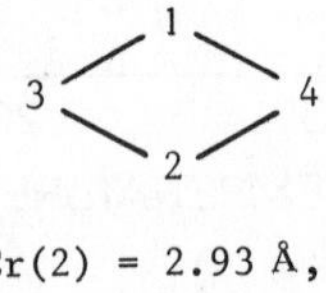

with Cr-Cr distances: Cr(1)-Cr(2) = 2.93 Å, Cr(1)-Cr(3) = 3.61 Å, and Cr(3)-Cr(4) = 6.55 Å. The structure is illustrated in Figure 4.17. Using

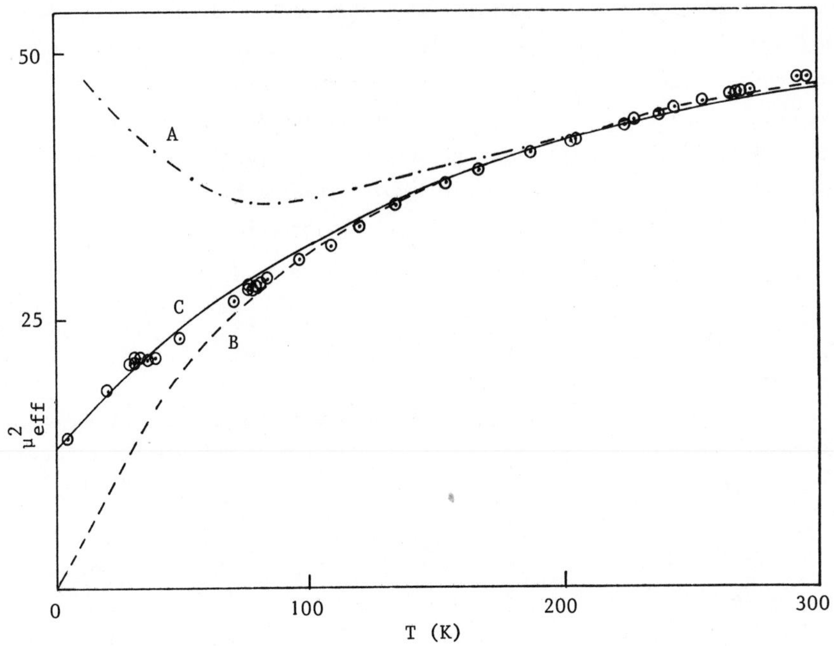

FIGURE 4.18 *Effective magnetic moment squared (calculated per tetranuclear ion) vs. temperature, calculated for a trigonal planar model (curve A, J/k = -20 K), tetrahedral model (curve B, J/k = -10 K), and planar-rhomboid model (curve C, J/k = -10.5 K, J_{12}/k = -20 K. Experimental data are indicated by the circles. From Ref. 44.*

a Hamiltonian of the form

$$\mathcal{H} = -2J[\vec{S}_1\cdot\vec{S}_3 + \vec{S}_2\cdot\vec{S}_3 + \vec{S}_1\cdot\vec{S}_4 + \vec{S}_2\cdot\vec{S}_4] - 2J'\vec{S}_1\cdot\vec{S}_2 \tag{4.13}$$

a fit to the powder susceptibility over a wide temperature interval was obtained (44), as illustrated in Figure 4.18, with parameters 2J/k of about -20 K and J'/k of about -40 K. The large distance between Cr(3) and Cr(4) was assumed to make any significant spin-spin interaction between these atoms unlikely, and the earlier analyses of the susceptibility data were made invalid because of the assumption of the wrong molecular geometry. Of especial interest is the fact that the higher order terms are not required in the Hamiltonian.

Subsequently the magnetic specific heat of $[Cr_4(OH)_6en_6](SO_4)_3\cdot 10H_2O$

was obtained (45). Two broad peaks were observed, at about 2.3 and 20 K. In Figure 4.19, we illustrate the specific heat behavior calculated (45) for several tetrameric models, and at least one broad peak is always obtained; in Figure 4.20, the specific heat for the planar rhomboid model, Eq. (4.13), is illustrated for a variety of ratios J/J'. The magnetic specific heat of the sulfate salt resembles these curves, but a final fit to the data was not obtained until the new interaction

$$\mathcal{H} = -2J''(\vec{S}_3 \cdot \vec{S}_4)$$

was also included. The sensitivity of the specific heat of this model to the new parameter J" is illustrated in Figure 4.21. The resulting best fit parameters are

$$2J/k = -22.8 \text{ K}$$
$$2J'/k = -42.6 \text{ K}$$
$$2J''/k = -7.6 \text{ K}$$

All parameters are antiferromagnetic in sign, and the major interactions are comparable in magnitude to those reported by Gray and co-workers. Sorai and Seki (45) also showed that the specific heat is more sensitive at low temperatures to the effect of the ratio J"/J' than is the susceptibility. Clearly, a complete magnetic study should always involve both isothermal susceptibility and heat capacity measurements, as well as complete structural information.

H. THE ISING MODEL

The Hamiltonian we have used so far for magnetic exchange is, as mentioned earlier, an isotropic one and is often called the Heisenberg model. We may rewrite

$$\mathcal{H} = -2J\, \vec{S}_1 \cdot \vec{S}_2$$

explicitly as

$$\mathcal{H} = -2J[\gamma(S_{1x}S_{2x} + S_{1y}S_{2y}) + S_{1z}S_{2z}] \tag{4.14}$$

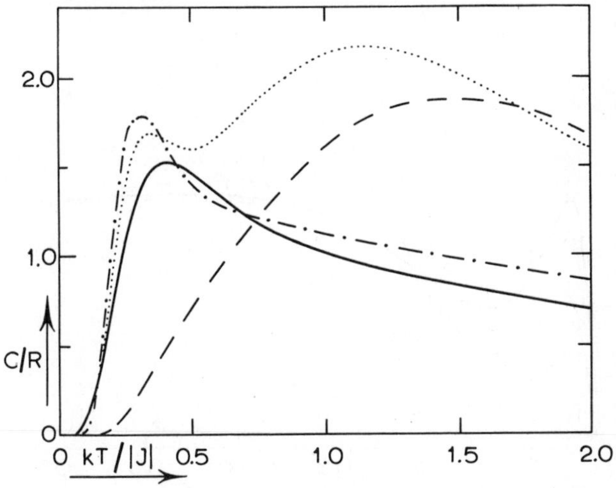

FIGURE 4.19 *The magnetic heat capacities associated with various models; — — — : trigonal planar; —.—.— : tetrahedral; ——— : planar-rhomboid, J'/J = 0.5; ..... : planar-rhomboid, J'/J = 1. From Ref. 45.*

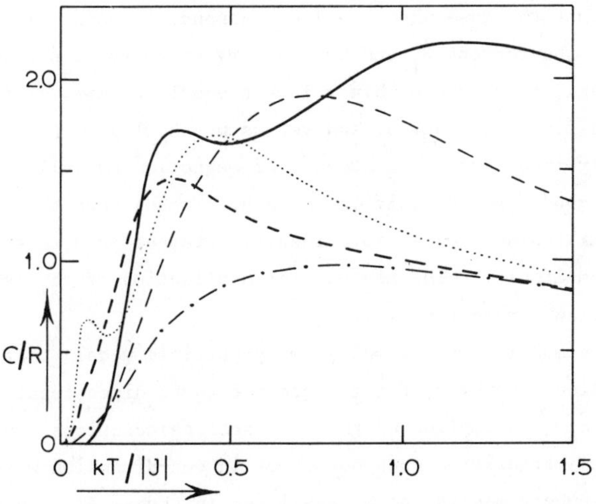

FIGURE 4.20 *The variation of the magnetic heat capacity arising from a planar-rhomboid model with the ratio J'/J; ——— : J'/J = 1; — — — : J'/J = 0.75; ..... : J'/J = 0.55; - - - : J'/J = 0.45; —.—.— : J'/J = 0.25. From Ref. 45.*

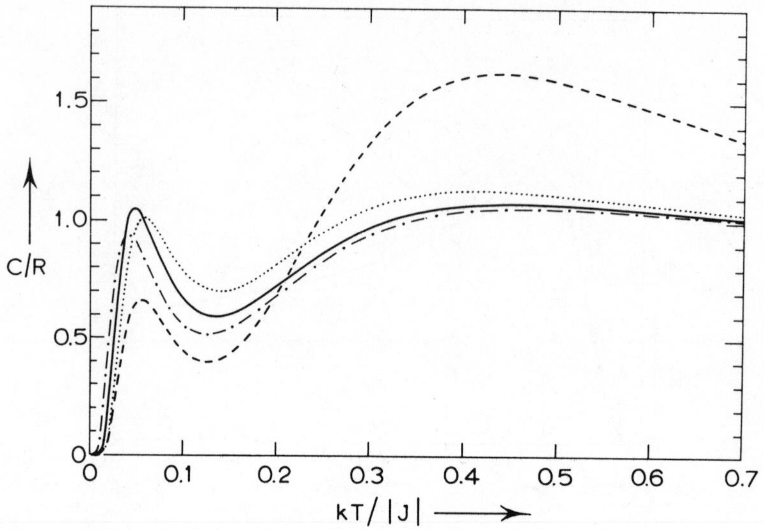

FIGURE 4.21 *The variation of the magnetic specific heat due to a planar-rhomboid model with the ratio J''/J;— — — : J''/J = 0; ..... : J''/J = 0.17; ——— : J''/J = 0.19; —.—.— : J''/J = 0.20. From Ref. 45.*

and observe that the case when $\gamma = 1$ corresponds to using all the spin-components of the vectors $\vec{S}_1$ and $\vec{S}_2$. A very anisotropic situation, called the Ising model, is obtained when γ is set equal to zero. This may appear as a very artificial situation, and yet as we shall see, the Ising model is exceedingly important in the theory of magnetism as well as in other physical many-body problems, if for no other reason than that solutions of the Ising Hamiltonian are far more readily obtained than those of the Heisenberg Hamiltonian. The history and application of the Ising model have been reviewed recently (46).

The Ising model is the simplest many-particle model that exhibits a phase transition. However, for an extended three dimensional lattice, the calculations are so complicated that no exact calculations have yet been done; the phase transition does not occur in one-dimension with either the Ising or Heisenberg models, as we shall see in Chapt. VI. In a two-dimensional lattice, the Ising model does exhibit a phase transition, and several physical properties may be calculated. Although it may not appear physically realistic, it will be useful to introduce the Ising model here and to illustrate the calculations by calculating the susceptibility of an isolated binuclear dimer.

The essence of the Ising model is that spins have only two orientations, either up or down with respect to some axis. We consider only the operator S_z, with eigenvalues + 1/2 and - 1/2 and so a dimer will exhibit but four states: + 1/2, + 1/2; + 1/2, - 1/2; - 1/2, + 1/2; and - 1/2, - 1/2. For the pair, we choose as the Hamiltonian

$$\mathcal{H} = -2JS_{z1}S_{z2} + g\mu_B H_z(S_{z1} + S_{z2}) \tag{4.15}$$

which has eigenvalues which may be obtained by inspection,

$$-(J/2 + g\mu_B H_z)$$
$$J/2 \text{ (twice)}$$
$$-(J/2 - g\mu_B H_z)$$

and we may write the partition function as

$$\begin{aligned} Z_a &= e^{-\langle\mathcal{H}\rangle/kT} \\ &= e^{(J/2+g\mu_B H_z)/kT} + 2e^{-J/2kT} + e^{(J/2-g\mu_B H_z)/kT} \\ &= 2e^{J/2kT}[e^{-J/kT} + \cosh(g\mu_B H_z/kT)] \end{aligned} \tag{4.16}$$

The molar magnetization M is defined as

$$M = NkT(\partial \ln Z_a/\partial H_z)_T \tag{4.17}$$

and thus, straightforward calculation leads to

$$M = \frac{Ng\mu_B \sinh(g\mu_B H_z/kT)}{e^{-J/kT} + \cosh(g\mu_B H_z/kT)} .$$

Since

$$\chi = (\partial M/\partial H_z)_T$$

we find

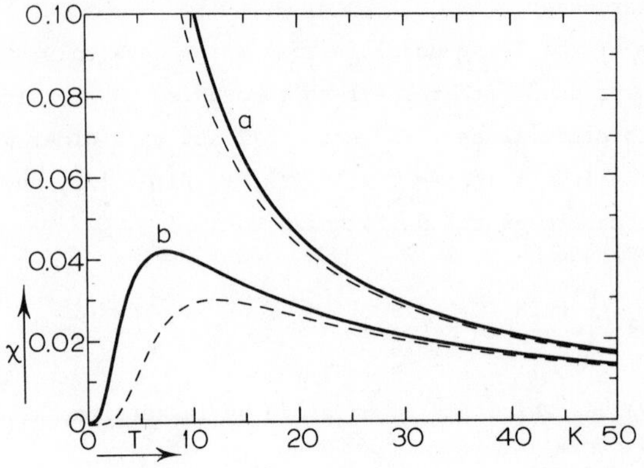

FIGURE 4.22 *Magnetic susceptibility of dimers of spin-1/2 ions coupled antiferromagnetically in the Heisenberg (dashed curves) and Ising (drawn curves) approximations (Eqs. (4.7) and (4.18)). |J/k| = 10 K with a : J > 0 and b : J < 0.*

$$\chi = \frac{2Ng^2\mu_B^2}{kT} \quad \frac{1 + e^{-J/kT}\cosh(g\mu_B H_z/kT)}{[e^{-J/kT} + \cosh(g\mu_B H_z/kT)]^2}$$

or in the limit of zero-field,

$$\chi_o = \frac{Ng^2\mu_B^2}{kT}\,(1 + e^{-J/kT})^{-1} \tag{4.18}$$

Note that this solution is actually only $\chi_\parallel$, the susceptibility parallel to the z-axis, and refers to a mole of dimers; an additional factor of 2 in the denominator is required in order to refer to a mole of ions. The behavior of Eq. (4.18) is compared with that of Eq. (4.7) in Figure 4.22.

The specific heat per mole of ions of the two-spin Ising system at zero field is easily calculated from Eq. (3.5) and Eq. (4.16), above, as

$$C = R(J/kT)^2\text{sech}^2(J/2kT) \tag{4.19}$$

and Eqs. (4.5) and (4.19) are compared in Figure 4.23. To date, there are no relevant data to compare this theory with experiment, although this model has been applied to the tetrameric compound cobalt acetylacetonate (47).

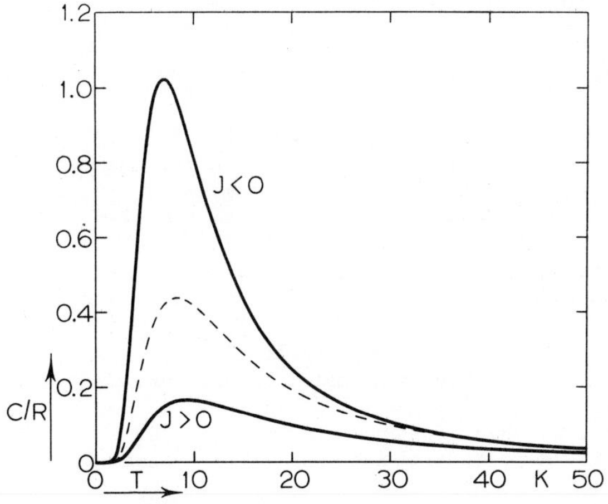

FIGURE 4.23 *Specific heat of dimers according to the Heisenberg (drawn curve) and Ising (dashed curve) approximations. (Eqs. (4.5) and (4.19)), $|J/k|$ = 10 K.*

REFERENCES:

1. R.L. Martin, "New Pathways in Inorganic Chemistry," E.A.V. Ebsworth, A.G. Maddock, and A.G. Sharpe, Eds., Cambridge University Press, London, 1968, Chapter 9.
2. P.W. Anderson, "Magnetism," (ed., G.T. Rado and H. Suhl), Vol. 1, p. 25, New York, Academic Press, Inc., 1963; K.W.H. Stevens, Phys. Reports (Phys. Lett. C) 24C, 1, February, 1976.
3. A. Carrington and A.D. McLachlan, "Introduction to Magnetic Resonance," Harper and Row, New York, 1967.
4. J.S. Smart, "Magnetism," (ed. G.T. Rado and Suhl), Vol. III, p. 63, New York: Academic Press, Inc., 1965.
5. S.A. Friedberg and C.A. Raquet, J. Appl. Phys. 39, 1132(1968).
6. B. Morosin, Acta Cryst. B26, 1203 (1970).
7. L. Berger, S.A. Friedberg, and J.T. Schriempf, Phys. Rev. 132, 1057 (1963).
8. A.P. Ginsberg, Inorg. Chim. Acta Reviews 5, 45(1971); D.J. Hodgson, Prog. Inorg. Chem. 19, 173 (1975); R.J. Doedens, Prog. Inorg. Chem. 21, 209 (1976).
9. B. Bleaney and K.D. Bowers, Proc. Roy. Soc. (London) A214, 451 (1952).
10. B.C. Guha, Proc. Roy. Soc. (London) A206, 353 (1951).
11. J.N. van Niekerk and F.R.L. Schoening, Acta Cryst. 6, 227 (1953).
12. P. de Meester, S.A. Fletcher, and A.C. Skapsky, J. Chem. Soc. - Dalton 1973, 2575.
13. G.M. Brown and R. Chidambaram, Acta Cryst. B29, 2393 (1973).
14. F.A. Cotton, B.G. De Boer, M.D. La Prade, J.R. Pipal and D.A. Ucko, Acta Cryst. B27, 1664 (1971).
15. G.F. Kokoszka, H.C. Allen, Jr., and G. Gordon, J. Chem. Phys. 42, 3693 (1965).

16. A. Abragam and B. Bleaney, "Electron Paramagnetic Resonance of Transition Ions," Oxford University Press, Oxford, 1970.
17. B.N. Figgis and R.L. Martin, J. Chem. Soc. 1956, 3837.
18. A.K. Gregson, R.L. Martin and S. Mitra, Proc. Roy. Soc. (London) A320, 473 (1971).
19. E. Kokot and R.L. Martin, Inorg. Chem. 3, 1303 (1964); L. Dubicki, C.M. Harris, E. Kokot and R.L. Martin, Inorg. Chem. 5, 93 (1966).
20. G.A. Barclay and C.H.L. Kennard, J. Chem. Soc. 1961, 5244.
21. A.K. Gregson and S. Mitra, J. Chem. Phys. 51, 5226 (1969); L.J. de Jongh and A.R. Miedema, Adv. Phys. 23, 1 (1974).
22. R. Whyman and W.E. Hatfield, Transition Metal Chemistry 5, 47 (1969).
23. R.C. Komson, A.T. McPhail, F.E. Mabbs, and J.K. Porter, J. Chem. Soc. A1971, 3447.
24. R.W. Jotham, J.C.S. Chem. Comm. 1973, 178.
25. D.B.W. Yawney, J.A. Moreland, and R.J. Doedens, J. Am. Chem. Soc. 95, 1164 (1973).
26. J.A. Moreland and R.J. Doedens, J. Am. Chem. Soc. 97, 508 (1974).
27. R.S.P. Coutts, R.L. Martin and P.C. Wailes, Aust. J. Chem. 26, 2101 (1973).
28. R. Jungst, D. Sekutowski, and G. Stucky, J. Am. Chem. Soc. 96, 8108 (1974).
29. A.P. Ginsberg, R.L. Martin, R.W. Brookes, and R.C. Sherwood, Inorg. Chem. 11, 2884 (1972).
30. J.R. Beswick and D.E. Dugdale, J. Phys. C (Solid State) 6, 3326 (1973); P.C. Benson and D.E. Dugdale, J. Phys. C (Solid State) 8, 3872 (1975).
31. A. Earnshaw and J. Lewis, J. Chem. Soc. 1961, 396.
32. O. Kahn and B. Briat, Chem. Phys. Lett. 32, 376 (1975).
33. E. Sinn, Coordin. Chem. Rev. 5, 313 (1970).
34. A.P. Ginsberg, R.L. Martin and R.C. Sherwood, Inorg. Chem. 7, 932 (1968).
35. B.N. Figgis and G.B. Robertson, Nature, Lond. 205, 694 (1965).
36. A. Earnshaw, B.N. Figgis and J. Lewis, J. Chem. Soc. (A) 1966, 1656 and references therein.
37. N. Uryû and S.A. Friedberg, Phys. Rev. 140, A1803 (1965).
38. M. Sorai, M. Tachiki, H. Suga and S. Seki, J. Phys. Soc. Japan 30, 750 (1971).
39. J. Ferguson and H.U. Güdel, Chem. Phys. Lett. 17, 547 (1972).
40. R.A.D. Wentworth and R. Saillant, Inorg. Chem. 6, 1436 (1967).
41. H. Kobayashi, I. Tsujikawa and I. Kimura, J. Phys. Soc. Japan 24, 1169 (1968).
42. C.G. Barraclough, H.B. Gray and L. Dubicki, Inorg. Chem. 7, 844 (1968).
43. M.T. Flood, R.E. Marsh and H.B. Gray, J. Am. Chem. Soc. 91, 193 (1969).
44. M.T. Flood, C.G. Barraclough and H.B. Gray, Inorg. Chem. 8, 1855 (1969).
45. M. Sorai and S. Seki, J. Phys. Soc. Japan 32, 382 (1972).
46. B.M. McCoy and T.T. Wu, "The Two Dimensional Ising Model," Harvard University Press, Cambridge, Mass., 1973.
47. J.C. Bonner, H. Kobayashi, I. Tsujikawa, Y. Nakamura, and S.A. Friedberg, J. Chem. Phys. 63, 19 (1975).

CHAPTER V

LONG-RANGE ORDER

A. INTRODUCTION

We now expand the concept of exchange to include interactions throughout a three-dimensional (3-d) crystalline lattice. The interactions are not necessarily long-range, and probably are important only for first through fourth nearest neighbors, but the effects are observed over large distances in a sample. Transitions to such long-range order are characterized by both characteristic specific heat anomalies and susceptibility behavior quite different from what has been described above. In other words, the transition to an ordered state is in fact a phase transition.

If the spins on a given lattice are all aligned spontaneously in the same direction, then the ordered state is a ferromagnetic one. No external field is required for this ordering, and in general an external field will destroy a ferromagnetic phase transition. The spontaneous ordering of spins persists below a certain critical (Curie) temperature, usually called T_c, and the susceptibility obeys a Curie-Weiss law at temperatures well above T_c, where the spins act as a paramagnetic system. One of the interesting facts here is that some of the phenomena associated with the phase transition, such as the specific heat anomaly, occur over exceedingly small temperature intervals.

B. MOLECULAR FIELD THEORY OF FERROMAGNETISM (1)

The magnitude of the spontaneous magnetization M is much larger than anything found with paramagnets. The classical or molecular field theory (MFT) assumes then that there is some sort of internal magnetic field H_m which orients the spins, while of course thermal agitation opposes this effect. The temperature T_c is that temperature at which the spontaneous magnetization loses the battle. An estimate of a typical H_m may be obtained by equating the two energies at T_c,

$$gS\mu_B H_m = kT_c,$$

and for metallic iron, $T_c \approx 1000$ K, $g \approx 2$ and $S \approx 1$, and so

$$H_m \approx 10^{-13}/2 \times 10^{-20} \approx 5 \times 10^6 \text{ Oersteds},$$

which is larger than any laboratory field. The value of H_m calculated is also larger than any dipole field, and requires the quantum concept of exchange in order to explain it.

Weiss introduced the MFT concept by assuming the existence of an internal field H_m which is proportional to the magnetization,

$$H_m = \lambda M \tag{5.1}$$

where λ is called the Weiss field constant. Restricting the discussion to temperatures above T_c for the moment, assume that the Curie law holds, but now the magnetic field, H_T, acting on the sample is the sum of H_m and the external field, H_{ext}.

$$H_T = H_{ext} + H_m \tag{5.2}$$

so

$$M/H_T = M/(H_{ext} + \lambda M) = C/T$$

Multiplying through,

$$M\,[\,1 - (\lambda C/T)] = CH_{ext}/T \tag{5.3}$$

or

$$\chi = M/H_{ext} = C/(T-\lambda C), \tag{5.4}$$

which is an exact result. When $H_{ext} = 0$, M is not zero at T_c (because of H_m) and so Eq. (5.3) yields under these conditions, $T_c = C\lambda$ and

$$\chi = C/(T-T_c) \tag{5.5}$$

which is the Curie-Weiss Law. Recall that Eq. (5.5) applies only above T_c, and that it is usually written as

$$\chi = C/(T-\theta) \tag{5.6}$$

which suggests that $\theta/T_c = 1$. While Eqs. (5.5) and (5.6) illustrate the close relationship between a non-zero Curie-Weiss θ and ferromagnetic exchange, few real materials obey the above relationship between θ and T_c.

Now let us turn to the spontaneous magnetization in the ferromagnetic region, $T < T_c$. Recall (Eq. (1.15)) that

$$M = Ng\mu_B J B_J(\eta)$$

with

$$\eta = g\mu_B H_T/kT,$$

and writing

$$H_T = H_{ext} + \lambda M,$$

then

$$\eta = g\mu_B(H_{ext} + \lambda M)/kT, \tag{5.7}$$

and let us examine the spontaneous behavior by setting $H_{ext} = 0$. Now, $\eta \to \infty$ as $T \to 0$, and recall that

$$B_J(\eta \to \infty) = 1$$

so that

$$M(T \to 0) = Ng\mu_B J,$$

the maximum M possible, and taking ratios,

$$M(T)/M(T = 0) = B_J(\eta) \tag{5.8}$$

But, we also have from Eq. (5.7) that

$$\eta = g\mu_B[\lambda M(T)]/kT$$

when $H_{ext} = 0$, or

$$M(T) = \eta kT/\lambda g\mu_B$$

which gives another expression for the ratio of M(T) and M(0),

$$M(T)/M(0) = \eta kT/Ng^2\mu_B^2 J\lambda \tag{5.9}$$

Eqs. (5.8) and (5.9) provide two independent relationships for M(T)/M(0), and so there must be two solutions at a given temperature. A trivial solution is obtained by setting the ratio equal to zero. The other solution is usually found graphically by plotting both ratios vs. η and finding the intersection. This method is sketched in Figure 5.1. Notice that for temperatures above T_c, the only intersection occurs at the origin which means that the spontaneous magnetization vanishes. This is consistent with the model. The curve for $T = T_c$ corresponds to a curve tangent to the Brillouin function at the origin, which is a critical temperature, while, at $T < T_c$, the intersections are at T = 0 and the point P.

Earlier we derived the relation

$$B_J(\eta) = [(J + 1)/3]\eta$$

for $\eta \ll 1$, and so the initial slope of $B_J(\eta)$ vs. η is (J + 1)/3. From Eq. (5.9) the initial slope is $kT/Ng^2\mu_B^2 J\lambda$ and so, equating the two initial slopes at $T = T_c$,

$$(J + 1)/3 = kT_c/Ng^2\mu_B^2 J\lambda,$$

we see that

$$T_c = Ng^2\mu_B^2 J(J + 1)\lambda/3k \tag{5.10}$$

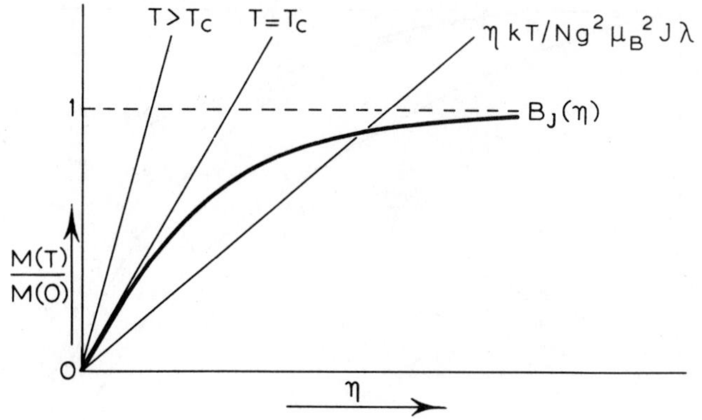

FIGURE 5.1 *Graphical method for the determination of the spontaneous magnetization at a temperature T.*

which predicts an increase in transition temperature with increasing total angular momentum and molecular field constant. Furthermore, substituting into Eq. (5.9), we have

$$M(T)/M(0) = (\eta kT/Ng^2\mu_B^2J)[Ng^2\mu_B^2J(J+1)]/3kT_c = [(J+1)/3](T/T_c)\eta \quad (5.11)$$

which is a universal curve, which should be applicable to all ferromagnets (1), and is illustrated in Figure 5.2 for several values of J.

Although the exact shape of this curve is somewhat in error (2), the general trend agrees with experiment, and in particular teaches us that the pretty pictures of totally aligned spins which we often draw are really applicable only as $T \to 0$.

Of especial interest are the two limits of T/T_c going to 0 or 1. First, as $T/T_c \to 0$, η becomes large and

$$M(T)/M(0) = B_J(\eta) = 1 - \frac{e^{-\eta}}{J} + \ldots..,$$

which, inserting Eq. (5.11) with $M(T)/M(0) = 1$, yields,

$$M(T)/M(0) = 1 - (1/J)\exp\{-[3/(J+1)](T_c/T)\} \quad (5.12)$$

The other limit, $T/T_c \to 1$, is the more interesting, in part because it is more accessible experimentally. Now, as T_c is approached from $T < T_c$, M(T)

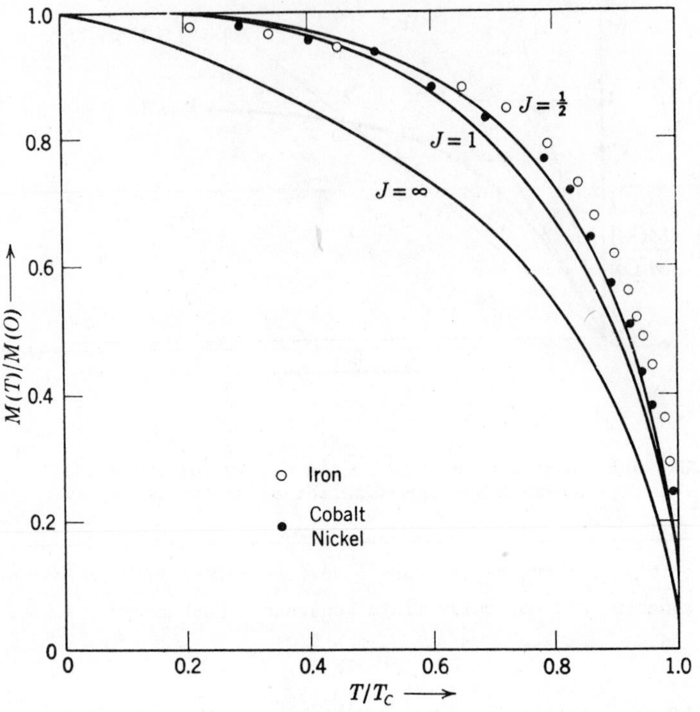

FIGURE 5.2 *The spontaneous magnetization as a function of temperature. The curves are obtained from theory; the points represent experimental data (F. Tyler, Phil. Mag. 11, 596 (1931)). From Ref. 1.*

is small, and

$$M(T)/M(0) \to 0,$$

or from Eq. (5.11),

$$M(T)/M(0) = [(J + 1)/3](T/T_c)\eta \to 0, \quad (5.13)$$

but since $(J + 1)/3$ is a number of the order of unity, as is also (T/T_c), then Eq. (5.13) requires that η must be very small in this situation. Thus,

$$M(T)/M(0) = B_J(\eta)$$

becomes, for small η,

$$M(T)/M(0) = \eta(J + 1)/3 - \eta^3[(J + 1)/3][(2J^2 + 2J + 1)/48] \qquad (5.14)$$

which is a more exact results than has been used heretofore. (Morrish (1) derives this equation with the numerical factor 48 replaced by 30). Since Eq. (5.11) provides another relationship between M(T)/M(0) and η, we may rearrange it in order to obtain

$$\eta = [M(T)/M(0)][3/(J + 1)](T_c/T),$$

and substituting into Eq. (5.14), obtain

$$[M(T)/M(0)]^2 = (10/3)\frac{(J + 1)^2}{J^2 + (J + 1)^2}[(T_c - 1)/T]$$

or

$$M(T)/M(0) = \left[(10/3)\frac{(J + 1)^2}{J^2 + (J + 1)^2}\right]^{\frac{1}{2}}(T_c/T - 1)^{\beta} \qquad (5.15)$$

where the classical molecular field derivation presented here requires a value of 1/2 for the exponent, β. It will be seen that the magnetization disappears continually as $T \to T_c$ from below, but with infinite slope at T_c.

The parameter β may be called a critical point exponent. The measurement of this (and related) parameters very close to T_c — with $\varepsilon \equiv (T_c/T - 1)$ approaching 10^{-5} or less in many cases — has been one of the most active areas recently in solid state physics (2). The principal reason for this is to compare experiment not with molecular field theory but with the extensive calculations done with both Ising and Heisenberg models of three-dimensional interactions; exact calculations cannot be carried out for this difficult problem, and so the interpretation of the critical point exponents has been helpful in determining the validity of the various approximation procedures. This is discussed further in Section F, below.

C. THERMAL EFFECTS

An ordered substance — whether it be a ferromagnet or an antiferromagnet — has complete spin order at (and only at) 0 K. As the temperature increases, increasing thermal energy competes with the exchange energy, causing a decrease in the magnetic order. Or, to put it another way, since

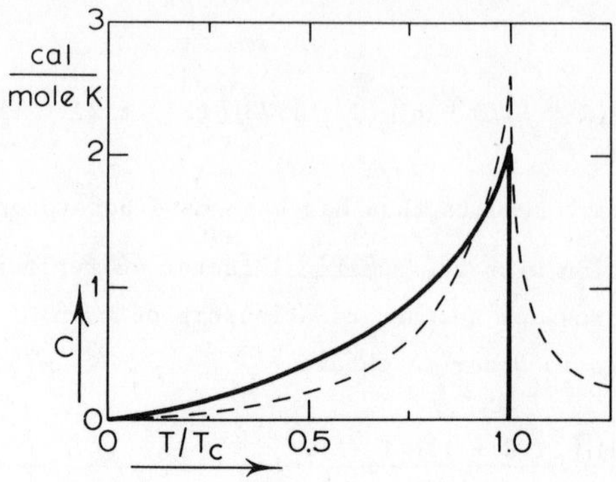

FIGURE 5.3 *The molecular-field variation of the magnetic specific heat with temperature (full line), compared with the measured values for nickel (broken line), to which the molecular-field results are scaled for the best fit. From Ref. 5.*

the Third Law requires the entropy of a substance to be zero at 0 K, as temperature increases the entropy of the magnetic system increases, and so there must be a magnetic contribution to the specific heat. The MFT uses a classical calculation of a magnet's self-energy in a field (1), a calculation which need not be repeated here. The result is as shown in Figure 5.3, the specific heat rises smoothly as T_c is approached from below and drops discontinously, as it must by MFT, at $T/T_c = 1$. A comparison of the magnetic specific heat of nickel with the MFT curve shows that, though the broad features are similar, that the experimental data rise more sharply ($T < T_c$) and fall more slowly ($T > T_c$). The magnetic specific heat of nickel above T_c is due primarily to short range order effects, presaging the onset of long-range order, and these effects are completely neglected by the classical theory. These results are typical of all long range order magnetic transitions.

D. MOLECULAR FIELD THEORY OF ANTIFERROMAGNETISM (1)

The exchange energy is sensitive to the spacing of the paramagnetic ions, and most magnetic insulators (i.e., the transition metal compounds which are the subject of this book) do not order ferromagnetically. Rather, neighboring spins are more frequently found to adopt an anti-parallel or antiferromagnetic arrangement. The antiferromagnetic transition likewise is a cooperative one, accompanied by a characteristic long-range

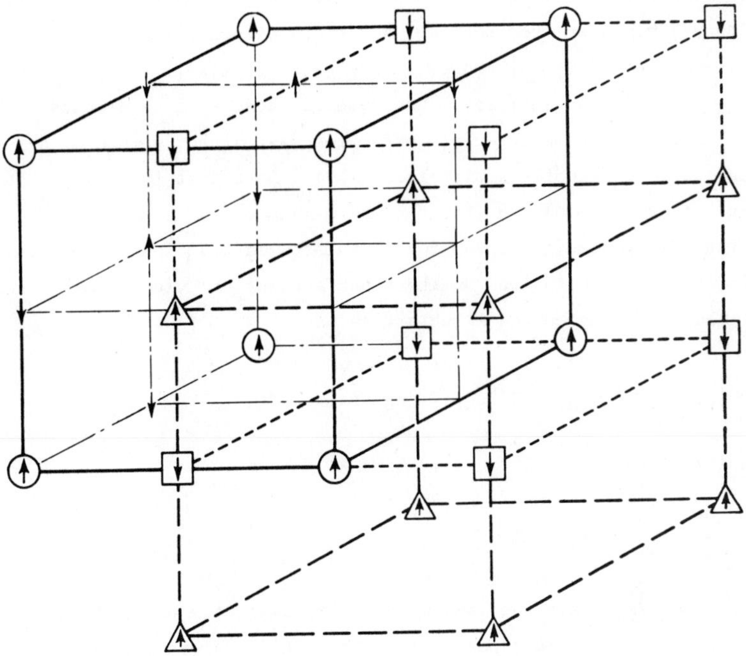

FIGURE 5.4 *Antiferromagnetism in a simple cubic lattice. The spins of the ions at the corners of the small cubes are arranged so that they form a series of interpenetrating cubic lattices with double the cell size. Three of these large cubes are shown, in heavy outline (ions denoted by circles), dashed line (ions denoted by triangles), dotted line (ions denoted by squares). The ions at the corners of any one large cube have all their spins parallel. From Ref. 4.*

ordering temperature which is usually called the Neel temperature, T_N. However, we adopt the admirable example set by de Jongh and Miedema (2) and use T_c as the abbreviation for a critical temperature, whether the transition be a ferromagnetic or antiferromagnetic one. This is in part acknowledgement that much of the theory in use is independent of the sign of the exchange, as well as because of the fact that, as we shall see, ferromagnetic interactions are often quite important for antiferromagnetic ordering

We may consider the usual antiferromagnet as consisting of two interpenetrating sublattices, with each sublattice uniformly magnetized with spins aligned parallel, but with the spins on one sublattice antiparallel to those on the other. A simple cubic lattice of this type is illustrated in Figure 5.4, taken from Ref. 4. A variety of other spin lattices have been discovered (5) — why should nature restrict itself to <u>two</u> interpenetrating magnetic sublattices? — and some of these will be discussed later. Neutron diffraction measurements offer the most direct method of determining the various magnetic sublattices (1,5). Note that this simple model requires that $M(T \rightarrow 0) = 0$ (for the whole sample).

The MFT is used as above, but it is assumed that ions on one lattice (A) interact only with ions on the B sublattice, and vice versa. Then, the field H_A acting on the A sublattice is

$$H_A = H_{ext} + H_{int\ B} = H_{ext} - \lambda M_B$$

while similarly,

$$H_B = H_{ext} - \lambda M_A$$

The negative signs are used because antiferromagnetic exchange effects tend to destroy alignment parallel to the field. Above T_c, the Curie Law is assumed, and one writes

$$M_A = \tfrac{1}{2}C(H_{ext} - \lambda M_B)/T$$
$$M_B = \tfrac{1}{2}C(H_{ext} - \lambda M_A)/T$$

and the total magnetization is $M = M_A + M_B$. Following procedures (1) similar to those used above, a modified Curie or Curie-Weiss law is derived,

$$\chi = C/(T + \theta'),$$

where θ' is $\lambda C/2$ in this case. Note the sign before the θ' constant, which is opposite to that found for ferromagnets. Again, MFT says that $T_c/\theta' \approx 1$, which is generally not true experimentally. Note also that any interaction with next-nearest- or other neighbors has been ignored.

It is probably worth pointing out that the common empirical use of the Curie-Weiss law is to write it as

$$\chi = C/(T - \theta) \tag{5.16}$$

and a plot of χ^{-1} vs. T yields a positive θ for ferromagnetic interactions, and a negative one for antiferromagnets.

On the other hand, it has been pointed out (6) that writing the Curie-Weiss law as

$$(\chi T)^{-1} = C^{-1}(1-\theta/T) \qquad (5.16a)$$

is a more convenient way of determining the Curie constant. A plot of $(\chi T)^{-1}$ vs. T^{-1} is a straight line (when $T >> \theta$) and yields directly the reciprocal of the Curie constant when we let $T^{-1} \to 0$. Furthermore, this procedure is a more sensitive indicator of the nature of the exchange interactions.

A list of a representative sample of antiferromagnets will be found in Table 5.1; more extensive tables are found in Refs. 7 and 8.

Now, an antiferromagnetic (AF) substance will follow the Curie-Weiss law above T_c, and if it is a cubic crystal, and if g-value anisotropy and zero-field splittings are unimportant, the susceptibility χ is expected to be isotropic. As T_c is approached from above, single-ion effects as well as short range order effects may begin to cause some anisotropy, and below T_c a distinctive anisotropy is required by theory and found experimentally. Thus, the following discussion concerns only measurements on oriented single crystals, and should not be confused with crystalline field anisotropy effects.

TABLE 5.1 Some Antiferromagnetic Substances[a]

Substance	T_c, K	Substance	T_c, K
$NiCl_2$	52	$CsMnBr_3 \cdot 2H_2O$	5.75[n]
$NiCl_2 \cdot 2H_2O$	7.258[b]	$Cs_2MnBr_4 \cdot 2H_2O$	2.82[n]
$NiCl_2 \cdot 4H_2O$	2.99[c]	$Rb_2MnBr_4 \cdot 2H_2O$	3.33[n]
$NiCl_2 \cdot 6H_2O$	5.34	$CsCoCl_3 \cdot 2H_2O$	3.38[g]
$MnCl_2$	1.96	$RbCoCl_3 \cdot 2H_2O$	2.975[h]
$MnCl_2 \cdot 2H_2O$	6.90[d]	$NiBr_2 \cdot 6H_2O$	8.30[i]
$MnCl_2 \cdot 4H_2O$	1.62	$FeCl_2 \cdot 4H_2O$	1.1
$MnCl_2 \cdot 4D_2O$	1.59	$K_3Fe(CN)_6$	0.129
$CoCl_2$	25	α-$RbMnCl_3 \cdot 2H_2O$	4.560[j]
CoF_2	37.7	Cs_3MnCl_5	0.601

TABLE 5.1 (Continued)

Substance	T_c, K	Substance	T_c, K
$CoCl_2 \cdot 2H_2O$	17.5	$[Ni\ en_3](NO_3)_2$	1.25[k]
$CoCl_2 \cdot 6H_2O$	2.29	$K_2Cu(SO_4)_2 \cdot 6H_2O$	0.029
$CuCl_2 \cdot 2H_2O$	4.3	$Ni[(NH_2)_2CS]_6Br_2$	2.25[l]
$Rb_2MnCl_4 \cdot 2H_2O$	2.24	$Co[(NH_2)_2CS]_4Cl_2$	0.92[m]
$Rb_2NiCl_4 \cdot 2H_2O$	4.65[f]	$Cs_3Cu_2Cl_7 \cdot 2H_2O$	1.62[o]
$Cs_2MnCl_4 \cdot 2H_2O$	1.81	$RbFeCl_3 \cdot 2H_2O$	11.96[p]
$MnBr_2 \cdot 4H_2O$	2.12	$CeCl_3$	0.345
$CoBr_2 \cdot 6H_2O$	3.150[e]	$GdCl_3$	2.20
$CoBr_2 \cdot 6D_2O$	3.225[e]	$Gd(OH)_3$	2.0
MnO	117	$GdCl_3 \cdot 6H_2O$	0.185
MnF_2	67.4	$ErCl_3$	0.307[q]
$CsMnCl_3 \cdot 2H_2O$	4.89	$ErCl_3 \cdot 6H_2O$	0.356[r]

a. Data are from Refs. 7 and 8, except where indicated.
b. L.G. Polgar, A. Herweijer, and W.J.M. de Jonge, Phys. Rev. B5, 1957 (1972).
c. J.N. McElearney, D.B. Losee, S. Merchant, and R.L. Carlin, Phys. Rev. B7, 3314 (1973).
d. J.N. McElearney, S. Merchant and R.L. Carlin, Inorg. Chem. 12, 906 (1973), and references therein.
e. K. Kopinga, P.W.M. Borm, and W.J.M. de Jonge, Phys. Rev. B10, 4690 (1974).
f. J.N. McElearney, H. Forstat, P.T. Bailey, and J.R. Ricks, Phys. Rev. B13, 1277 (1976).
g. A. Herweijer, W.J.M. de Jonge, A.C. Botterman, A.L.M. Bongaarts, and J.A. Cowen, Phys. Rev. B5, 4618 (1972).
h. J. Flokstra, G.J. Gerritsma, B. van den Brandt, and L.C. van der Marel, Phys. Lett. 53A, 159 (1975); J.N. McElearney and S. Merchant, unpublished.
i. S.N. Bhatia and R.L. Carlin, Physica 86-88B, 903 (1977).
j. W.J.M. de Jonge and C.H.W. Swüste, J. Chem. Phys. 61, 4981 (1974).
k. A.J. van Duyneveldt, A. van der Bilt, J.P.C. Vreugdenhil, and R.L. Carlin, Chem. Phys. Lett. 26, 100 (1974).
l. H. Forstat, N.D. Love, and J.N. McElearney, J. Chem. Phys. 43, 1626 (1965).
m. H. Forstat, N.D. Love, J.N. McElearney, and H. Weinstock, Phys. Rev. 145, 406 (1966).
n. C.H.W. Swüste, W.J.M. de Jonge, and J.A.G.W. van Meijel, Physica 76, 21 (1974).
o. T.O. Klaassen, S. Wittekoek, J.J. van der Klink, and N.J. Poulis, Physica 41, 523 (1969).
p. K. Kopinga, Thesis, Eindhoven, 1976.
q. C.W. Fairall, J.A. Cowen, and E. Grabowski, Phys. Lett 35A, 405 (1971).
r. E. Lagendijk and W.J. Huiskamp, Physica 65, 118 (1973).

Imagine a perfectly antiferromagnetically aligned crystal at 0 K. There is some crystallographic direction, to be determined only by experiment and called the preferred or easy axis, along which the spins are aligned. The easy axis may or may not coincide with a crystallographic axis. A small external magnetic field, applied parallel to the easy axis, can cause no torque on the spins, and since the spins on the oppositely-aligned sublattices cancel each other's magnetization, then $\chi_{\parallel} \rightarrow 0$ as $T \rightarrow 0$. Note that "parallel" in this case denotes only the relative orientation of the axis of quantization of the aligned spins and the external field. As

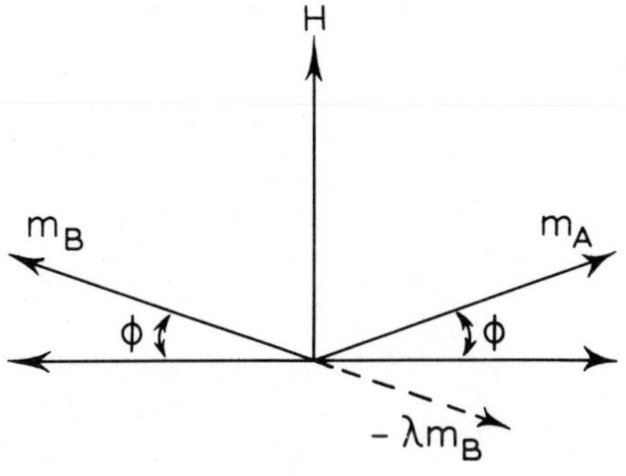

FIGURE 5.5 *The magnetization of an antiferromagnet when the field is applied at right angles to the spin orientation. Each sublattice rotates through a small angle* ϕ, *yielding a net magnetic moment.*

the temperature rises, the spin alignment is upset as usual by thermal agitation, the external field tends to cause some torque, and an increasing $\chi_{\parallel}$ is observed.

A field applied perpendicular to the easy axis tends to cause spins to line up, but (Fig. 5.5) the net resulting couple on each pair of dipoles should be zero. The net result of this effect is that $\chi_{\perp}$ remains approximately constant below T_c. A typical set of data (4) is illustrated in Figure 5.6, where the described anisotropy is quite apparent. Notice also that the susceptibility of a powder is described as

$$\chi_{\text{powder}} = \langle\chi\rangle = \tfrac{1}{3}(\chi_{\parallel} + 2\chi_{\perp}) \tag{5.17}$$

and so $\langle\chi\rangle$ at 0 K is 2/3 its value at T_c. The MFT also yields (9) two other equations of some interest.

$$\theta = 2S(S + 1)zJ/3k \tag{5.18}$$

and, in the limit of zero-field for a uniaxial antiferromagnet,

$$\chi_{\perp}(T=0) = Ng^2\mu_B^2/4zJ \tag{5.19}$$

In Eq. (5.18), θ is the Curie-Weiss constant, S is the spin, J/k the exchange constant in Kelvins, and z the coordination number of a lattice site. The parameters in Eq. (5.19) have been defined, and the equation applies to $\chi_{\perp}$ at T = 0 K. Both relationships show the intimate correlation between the observable parameters and the exchange constant.

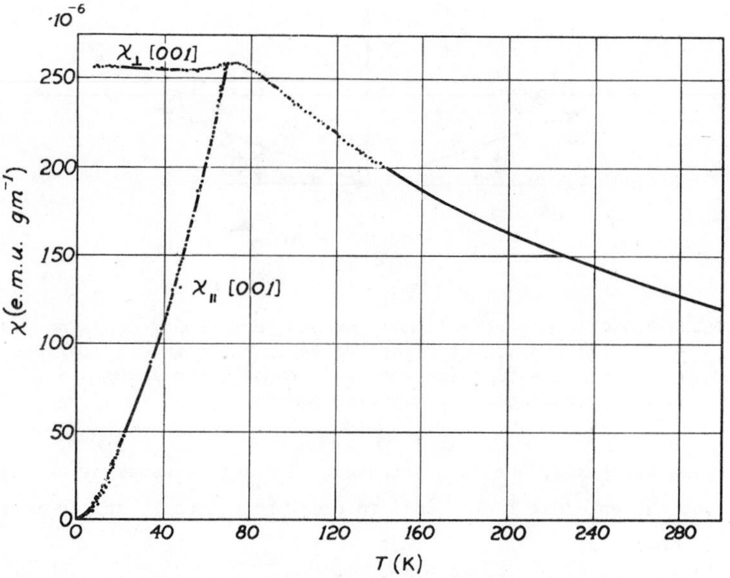

FIGURE 5.6 *The susceptibility of MnF_2 parallel and perpendicular to the [001] axis of the crystal (data of S. Foner). From Ref. 4.*

It is important to distinguish the ordering temperature, T_c, and the temperature at which a maximum occurs in the susceptibilities. They are not the same. Fisher (10) presented an argument that the temperature variation of the specific heat of an antiferromagnet is essentially the same as that of the temperature derivative of the susceptibility. He established the relation

$$C(T) = A(\partial/\partial T)[T\chi_{\parallel}(T)] \tag{5.20}$$

where the constant of proportionality A is a slowly varying function of temperature. This expression implies that any specific heat anomaly will be associated with a similar anomaly in $\partial(\chi_{\parallel}T)/\partial T$. Thus the specific heat

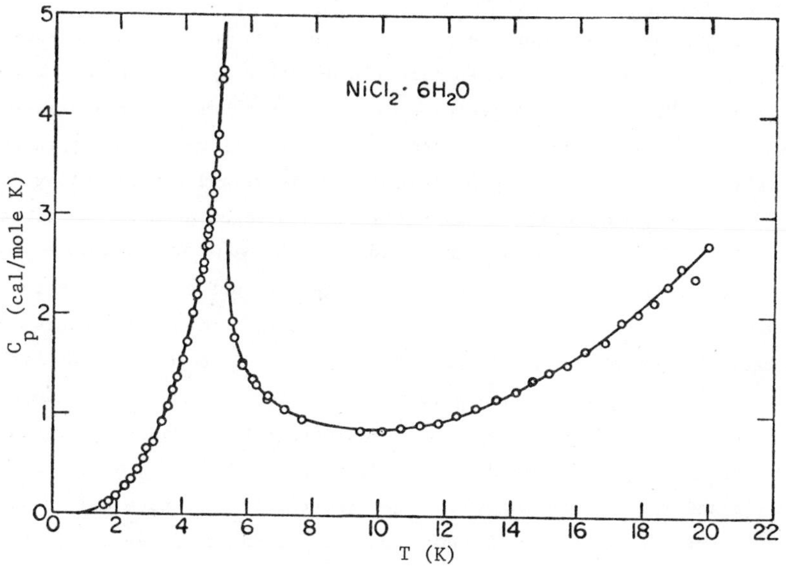

FIGURE 5.7 *Specific heat of $NiCl_2.6H_2O$. From Ref. 12.*

singularity (λ-type anomaly) normally observed at the antiferromagnetic transition temperature T_c (see below) is associated with a positively infinite <u>gradient</u> in the susceptibility at T_c. The maximum in $\chi_{\parallel}$ which is observed experimentally in the transition region must lie somewhat <u>above</u> the actual ordering temperature, and the assignment of T_c to T_{max} by the MFT is due, once again, to the exclusion of short-range order effects. The exact form of Eq. (5.20) has in fact been substantiated by careful measurements (11) on $CoCl_2.6H_2O$, and the validity of the argument has been checked repeatedly.

As with ferromagnetism, the onset of antiferromagnetism likewise causes a sharp anomaly in the magnetic specific heat. It is frequently λ-shaped and occurs over a small temperature interval. A typical set of data, in this case for $NiCl_2.6H_2O$, is illustrated in Figure 5.7. The observation of such a λ-shaped curve is adequate proof that a phase transition has occurred, but of itself cannot distinguish whether the transition is to an antiferromagnetic or ferromagnetic state, or whether the transition is in fact magnetic in origin. Further experiments, such as susceptibility measurements, as well as the specific heat in a non-zero external magnetic field, are required before the correct nature of the phase transition may become known.

This may be an appropriate place to point out that there are some qualitative correlations between magnetic dilution and T_c that are interesting. Anhydrous metal compounds order at higher temperatures, generally, than do the hydrates; the metal atoms are usually closer together in compounds such as $NiCl_2$ than in the several hydrates, and water usually does not furnish as good superexchange paths as do halogen atoms. The first four entries in Table 5.1 are in accord with the suggestion, though the lack of an exact trend is clear. The ordering temperature will clearly increase with increasing exchange interaction, but there are no simple models available to suggest why, for example, T_c for $NiCl_2.6H_2O$ should be higher than that for $NiCl_2.4H_2O$. The determining factor is the superexchange interaction, and this depends on such factors as the ligands separating the metal atoms, the distances involved, as well as the angles of the metal-ligand-metal exchange path. The reader will also note the unusually low transition temperature of $MnCl_2$, as well as the relative unimportance of the presence of water of hydration upon T_c for the trihalides of the lanthanides.

The molar entropy change associated with any long-range spin ordering is always $R\ln(2S+1)$, where we may ignore Schottky terms for the moment. Once the magnetic specific heat is known, it may be integrated in the usual fashion in order to obtain the magnetic entropy. The full $R\ln(2S+1)$ of entropy is never acquired between $0 < T < T_c$ because, although long-range order persists below T_c, short-range order effects always contribute above T_c. Calculations (9) for spin-1/2 face-centered cubic lattices in either the Ising or Heisenberg model, for example, show that, respectively 14.7 and 38.2 % of the total entropy must be obtained above T_c. As an example, the entropy change (12) for magnetic ordering in $NiCl_2.6H_2O$ is illustrated in Figure 5.8; only 60 % of the entropy is acquired below T_c.

As was mentioned in Chapt. II, the magnetic ordering specific heat contribution follows a T^{-2} high temperature behavior. In fact (9), the relationship is

$$CT^2/R = 2S^2(S + 1)^2 zJ^2/3k^2 \tag{5.21}$$

which again allows an estimation of the exchange parameter, J/k.

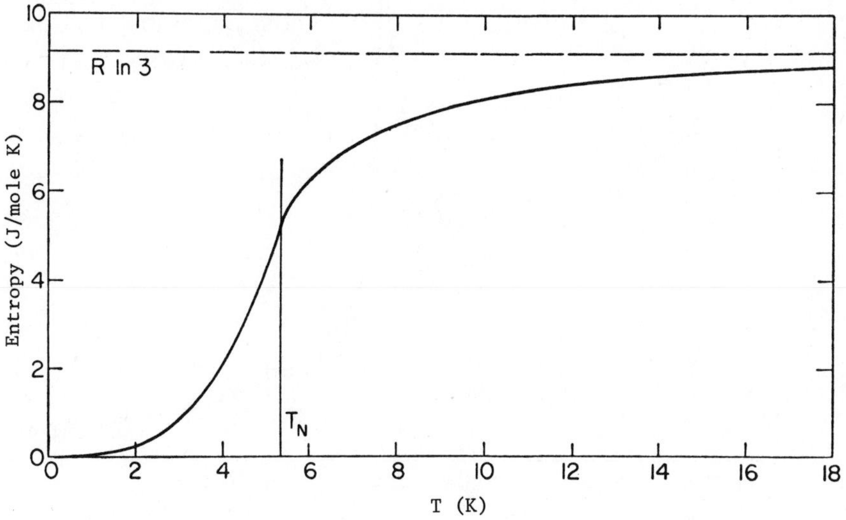

FIGURE 5.8 *Magnetic entropy as a function of temperature for $NiCl_2 . 6H_2O$. From Ref. 12.*

E. ISING, XY, AND HEISENBERG MODELS (2,9)

The molecular field theory offers a remarkably good approximation to many of the properties of three-dimensionally ordered substances, the principal problem being that it neglects short-range order. As short-range correlations become more important, say as the dimensionality is reduced to two or one, molecular field theory as expected becomes a worse approximation of the truth. This will be explored in the next chapter.

In order to discuss exchange, one must introduce quantum-mechanical ideas, and as has been alluded to already, we require the Hamiltonian

$$\mathcal{H} = -2J \sum_{\substack{i,j \\ i \neq j}} \vec{S}_i \cdot \vec{S}_j \tag{5.22}$$

The mathematical complexities of finding an exact solution of this Hamiltonian for a three-dimensional lattice so as to predict, for example,

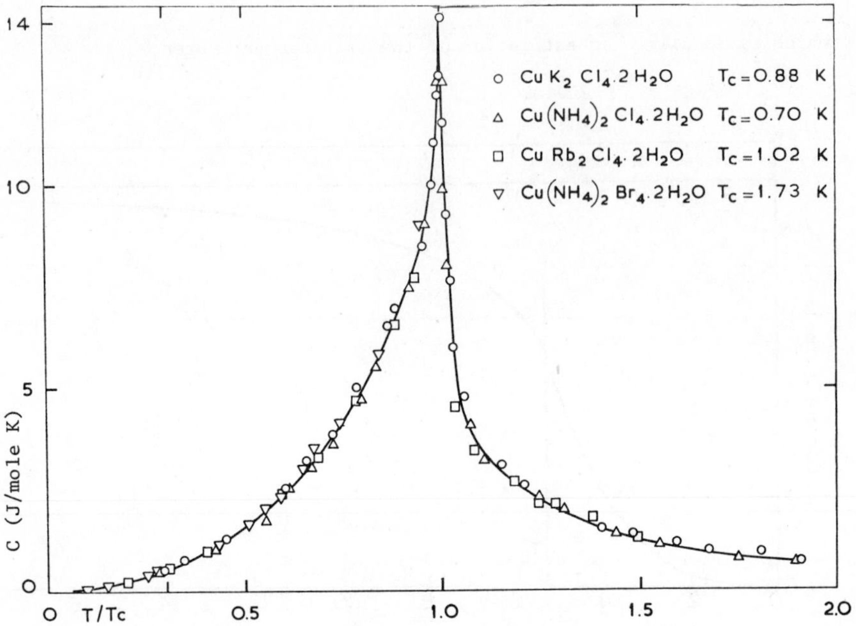

FIGURE 5.9 *Heat capacities of four isomorphous ferromagnetic copper salts (S = 1/2). From Ref. 2.*

the shape of the specific heat curve near the phase transition, are enormous, and have to date prevented such a calculation. On the other hand, the careful blending of experiment with numerical calculations has surely provided most of the properties we require (2). For example, the specific heats of four isomorphic copper salts, all ferromagnets, are plotted on a universal curve in Figure 5.9. It is clear (2,13) that the common curve is a good approximation of the body centered cubic (b.c.c.) Heisenberg ferromagnet, S = 1/2, whose interactions are primarily nearest neighbor.

The first approximation usually introduced in finding solutions of Eq. (5.22) is to limit the distance of the interactions, most commonly to nearest-neighbors. The dimensionality of the lattice is also of special importance, as will be discussed in the next chapter; in particular, the behavior of the various thermodynamic quantities changes more between changes of the lattice dimensionality (1, 2, or 3) than they do between different structures (say, simple cubic, face-centered cubic, body-centered cubic) of the same dimensionality. The last choice to be made, and again a significant one, is the choice of magnetic model or approximation to be investigated. There are three limiting cases that have been extensively explored. Two of these, the Ising and Heisenberg models, have been introduced already; the third, the XY model, is also of some importance, and is obtained by setting $\gamma = 1$ in Eq. (4.16) and letting the S_z contribution be zero. A number of intermediate situations can also be visualized.

Isotropic interactions are required in order to apply the Heisenberg model, and this suggests first of all that the metal ions should reside in sites of high symmetry. In view of all the possible sources of anisotropy — thermal contraction on cooling, inequivalent ligands, low symmetry lattice, etc.— it is surprising that there are a number of systems which do exhibit Heisenberg behavior. Clearly, single ion anisotropy, whether it be caused by zero-field splittings or g-value anisotropy, must be small, and this is why the S-state ions Mn^{2+}, Fe^{3+}, Gd^{3+}, and Eu^{2+}, offer the most likely sources of Heisenberg systems. Recall that g-value anisotropy is related to spin-orbit coupling, which is zero, to first order, for these ions. Copper(II) offers, to date, the next best examples of Heisenberg systems that have been reported because the orbital contribution is largely quenched. Being an $S = 1/2$ system, there are no zero field splittings to complicate the situation, and the g-value anisotropy is typically not large. For example, $g_{\parallel} = 2.38$, $g_{\perp} = 2.06$ for $K_2CuCl_4 \cdot 2H_2O$ (2,14). Trivalent chromium and nickel(II) are also potentially Heisenberg ions, though only in those compounds where the zero-field splittings are small compared to the exchange interactions.

Ising ions require large anisotropy, and since the magnetic moment varies with the g-value, it is possible to take g-value anisotropy as a reliable guide to finding Ising ions. The large anisotropy in g-values of octahedral cobalt(II) was mentioned in Chapt. III, and this ion provides some of the best examples of Ising systems. Similarly, tetrahedral cobalt(II) can be highly anisotropic when the zero-field splitting is large, and forms a number of Ising systems. Thus, Cs_3CoCl_5 has $2D/k = -12.4$ K, and with $T_c = 0.52$ K, this compound follows the Ising model (15). On the other hand, if the zero-field splitting of tetrahedral cobalt(II) is small compared to the exchange interactions, then it should be a Heisenberg ion! If, however, the zero-field splitting is large and positive, then tetrahedral cobalt(II) may be an XY-type ion. This will be discussed in more detail in Chapter VIII.

Because of zero-field splittings, Dy^{3+} is also frequently an Ising ion (2). The fact of anisotropy seems to be more important than the cause of it in a particular sample.

The concept of "long-range order" may not be defined precisely, at least in an operational sense. The term is usually restricted to 3-d order existing over more than a few atoms. As we have seen, it is indicated by a λ-anomaly in the specific heat at the critical temperature, but we can not say how long, "long-range" order is. Consider the following data, for example. Magnetic impurities often exhibit their own characteristic features even when they are quite dilute. The specific heat (16) of $KMnCl_3.2H_2O$ is illustrated in Figure 5.10. The major peak indicates magnetic ordering at 2.70 K, but notice also the minor peak at 1.62 K. This

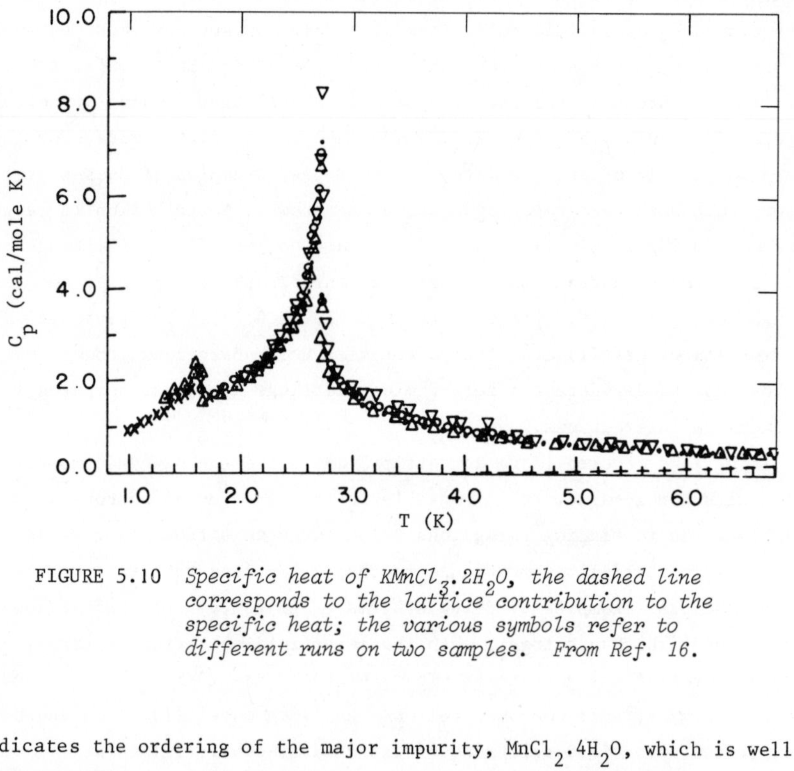

FIGURE 5.10 *Specific heat of $KMnCl_3.2H_2O$, the dashed line corresponds to the lattice contribution to the specific heat; the various symbols refer to different runs on two samples. From Ref. 16.*

indicates the ordering of the major impurity, $MnCl_2.4H_2O$, which is well known to order at 1.62 K. One may estimate that the tetrahydrate is present at only a few percent and yet it continues to exhibit its own characteristic behavior. It is not known whether the impurity clusters, or is distributed evenly throughout the crystal.

Table 5.2, taken from the extensive article (2) of de Jongh and Miedema, reports some of the critical properties of 3-d Ising models and

TABLE 5.2 Critical entropy parameters of theoretical 3-d Ising models, $S = 1/2$, and their experimental approximants (2). The numbers in parentheses refer to the number of nearest magnetic neighbors.

Compound or model	J/k(K)	T_c(K)	T_c/θ	S_c/R	$(S_\infty - S_c)/R$	$(S_\infty - S_c)/S_c$
Ising, diamond (4)			0.6760	0.511	0.182	0.356
Ising, s.c. (6)			0.7518	0.5579	0.1352	0.2424
Ising, b.c.c. (8)			0.7942	0.5820	0.1111	0.1909
Ising, f.c.c. (12)			0.8163	0.5902	0.1029	0.1744
MFT (∞)			1.000	0.693	0.000	0.000
Rb_3CoCl_5	-0.511	1.14	0.74	0.563	0.137	0.24
Cs_3CoCl_5	-0.222	0.52	0.79	0.593	0.106	0.18
$DyPO_4$	-2.50	3.390	0.678	0.505	0.185	0.37
$Dy_3Al_5O_{12}$	-1.85	2.54	0.68	0.489	0.204	0.42
$DyAlO_3$	≈ -2	3.52	0.62	0.521	0.172	0.33

compounds. The total entropy to be acquired by magnetic ordering for spin-1/2 systems is $S_\infty = 0.693R$, and S_c indicates the amount of entropy acquired below T_c. One fact immediately apparent from the calculations is that some 15-25 % of the entropy must appear as short-range entropy, above T_c.

The isomorphous compounds Cs_3CoCl_5 and Rb_3CoCl_5 provide two of the best examples of three-dimensional Ising systems (2,15,17). The structure (18), in Figure 5.11, consists of isolated tetrahedral CoX_4^{2-} units, along with extra cesium and chloride ions. All the magnetic ions are equivalent, and although the crystal is tetragonal, each Co ion has six nearest Co neighbors in a predominantly simple cubic environment. As has been pointed out above, 2D/k is large and negative in the cesium compound, and appears (17) to be negative and even larger in magnitude in the rubidium compound. Thus, only the $S_z = \pm 3/2$ states need be considered, $g_\perp = 0$, and the two systems meet the requirements of being Ising lattices with effective spin-1/2. The ordering temperatures are 0.52 K (Cs) and 1.14 K (Rb), and the specific heats of both systems are plotted in Figure 5.12 on a reduced

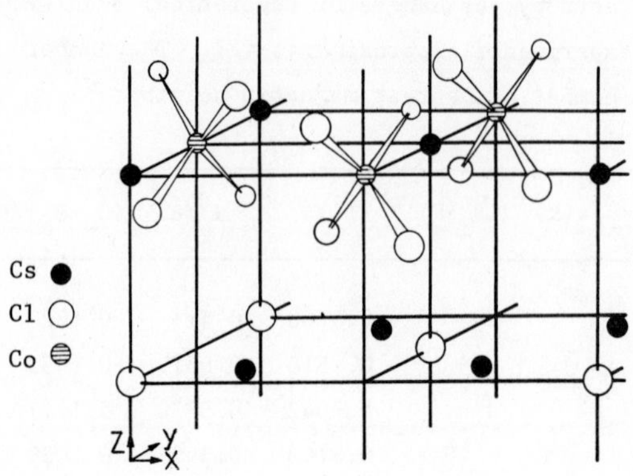

FIGURE 5.11 *Crystal structure of Cs_3CoCl_5. From Ref. 18.*

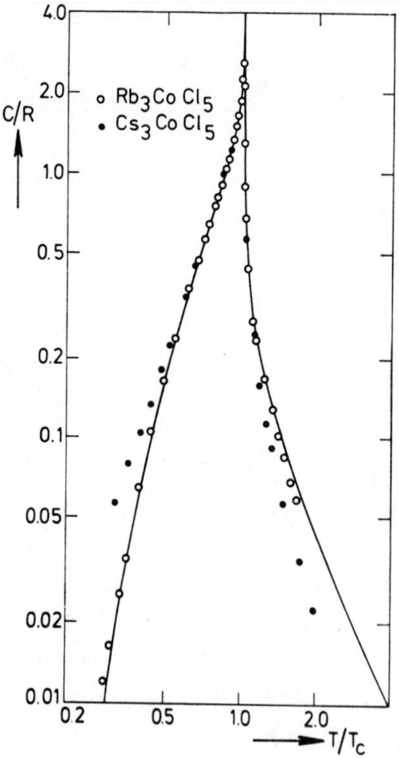

FIGURE 5.12 *Heat capacities of $CoRb_3Cl_5$ and $CoCs_3Cl_5$ compared with the theoretical prediction for the simple cubic Ising model. From Ref. 2.*

scale. (The Schottky contribution lies at much higher temperatures, and the lattice contribution is essentially zero at these temperatures.) The curve is a calculated one, based on series expansion techniques, for a 3-d Ising lattice, and perfect agreement is obtained for the rubidium system. The lower ordering temperature of the cesium compound causes dipole-dipole interactions (Section H) to be more important, which probably causes the small discrepancies; another way of putting this is that in the cesium compound, a larger effective number of neighbors must be assumed, which suggests that it may be closer to a b.c.c. Ising system.

Let us turn now to the Heisenberg model, in Table 5.3, which is also taken from de Jongh and Miedema (2). Calculations for the Heisenberg model are more difficult than those for the Ising model, and thus only $S = \frac{1}{2}$ and $S = \infty$ are listed, the latter case corresponding to a classical spin. The examples listed are those whose anisotropy amounts to 1 % or less. Notice immediately that S_c/R is smaller for the Heisenberg model than for the Ising model, which says that short-range ordering is more important in the Heisenberg system. The ratio T_c/θ also indicates this. Furthermore, short-range order effects are also enhanced by the lowering of the spin, S.

From the experimentalists' point of view, the principal problem with the Heisenberg model is to find materials that are sufficiently isotropic to warrant application of the available theory. A cubic crystal structure, for example, is obviously a preferable prerequisite, yet a material that is of high symmetry at room temperature, which is where most crystallographic work is done, may undergo a crystallographic phase transition on going to the temperature where the magnetic effects become observable. Single ion anisotropies must of course be as small as possible, and in this regard it is of interest that $KNiF_3$ is a good example of a Heisenberg system. With a spin-1, zero-field splittings may be anticipated, but in this particular case the salt assumes a cubic perovskite structure and exhibits very small anisotropy. The high transition temperature, 246 K, on the other hand, causes difficulty in the accurate determination of such quantities as the magnetic specific heat. The compound $RbMnF_3$ is isomorphous and also isotropic but then the high spin (5/2) makes difficult the theoretical calculations for comparison with experimental data.

The specific heat of four Heisenberg ferromagnets has already been referred to and shown in Figure 5.9. The compounds have tetragonal unit cells, but the c/a ratios are close to 1. The difficulty in applying the Heisenberg model here lies with the problem that nearest-neighbor inter-

TABLE 5.3 Critical entropy parameters of theoretical 3-d Heisenberg models. The nearest-neighbor $S = \frac{1}{2}$ and $S = \infty$ models are listed. The values refer to ferromagnets. In case of T_c/θ the values for antiferromagnets ($S=\frac{1}{2}$) have been added (minus sign). For references to the experimental data see the text or (1,2).

Model or compound	S	T_c(K)	J/k(K)	T_c/θ	S_c/R	$(S_\infty - S_c)/R$	$(S_\infty - S_c)/S_c$
Heisenberg, s.c. (z=6)	1/2			0.56(+) 0.64(−)	0.43	0.26	0.60
Heisenberg, b.c.c. (z=8)	1/2			0.63(+) 0.70(−)	0.45	0.24	0.53
Heisenberg, f.c.c. (z=12)	1/2			0.67(+) 0.72(−)	0.46	0.23	0.50
Heisenberg, s.c. (z=6)	∞			0.72(+)		0.42	
Heisenberg, b.c.c. (z=8)	∞			0.77(+)		0.34	
Heisenberg, f.c.c. (z=12)	∞			0.79(+)		0.31	
Ferromagnets							
$(NH_4)_2CuCl_4.2H_2O$	1/2	0.701	0.23	0.77			
$K_2CuCl_4.2H_2O$	1/2	0.877	0.30	0.74			
$Rb_2CuCl_4.2H_2O$	1/2	1.02			0.47	0.22	0.46
$(NH_4)_2CuBr_4.2H_2O$	1/2	1.83	0.60	0.74			
$Rb_2CuBr_4.2H_2O$	1/2	1.87	0.63	0.74			
EuO	7/2	69		0.80			
EuS	7/2	16.4		0.81			
Antiferromagnets							
$CuCl_2.2H_2O$	1/2	4.36	?	?	0.43	0.23	0.54
NdGaG	1/2	0.516	≈ −0.34	≈ 0.76	0.46	0.26	0.58
SmGaG	1/2	0.967	≈ −0.60	≈ 0.81	0.42	0.27	0.65
$KNiF_3$	1	246	−44	0.72			
$RbMnF_3$	5/2	83.0	−3.40	0.70			
MnF_2	5/2	67.33	−1.76	0.79	1.53	0.26	0.17

actions alone do not fit the data; rather, second neighbors also had to be included. In fact, recent work (19) suggests that as many as seventeen equivalently interacting magnetic neighbors need to be considered.

The compound $CuCl_2 \cdot 2H_2O$ serves as a good example of the 3-d Heisenberg S = 1/2 antiferromagnet, and is of further interest because so many other copper compounds seem to be ferromagnetic. The crystal has a chain-like structure of copper atoms bridged by two chlorine atoms, and this has led some investigators to try to interpret the properties of the compound as if it were a magnetic chain as well (cf. Chapt. VI). For example, 33 % of the spin entropy appears above T_c. De Jongh and Miedema (2) have argued, however, that pronounced short-range order is characteristic of a 3-d Heisenberg magnet, and that theory predicts in this case as much as 38 % of the entropy should be obtained above T_c.

The only 3-d, XY antiferromagnets known to date are the S = 1/2 systems, $[Co(C_5H_5NO)_6](ClO_4)_2$ and $[Co(C_5H_5NO)_6](BF_4)_2$. These are discussed in Chapt. VIII.

F. CRITICAL POINT EXPONENTS (2,3,20)

This subject was introduced above in Section B, and has been the subject of much recent work in both theoretical and experimental solid state physics. For our purposes in this book, it will be enough to call the work to the attention of the reader and suggest a rationale for it. At the least, it is a tour de force of experimental technique, for data points have been taken, especially in heat capacity experiments, with microdegree resolution. The experimental variable of interest is $\varepsilon = 1 - T/T_c$, with different behavior usually being observed as T is above or below T_c. Some data for two manganese salts are illustrated in Figure 5.13.

The reason behind careful studies such as these is that one does not enter the critical region until the transition point is approached closely enough. The critical region is that temperature range around T_c in which the behavior of the thermodynamic functions appears to be governed by the power-law expressions which have been developed. It is confined to a region $|T-T_c|/T_c < 10^{-1}$ or 10^{-2}, typically, and experiments must of course lie within this range in order to test the theoretical models. The critical point exponents depend on the nature of the interactions and the dimension of the lattice, among other things, and have been calculated for such functions as the specific heat, spontaneous magnetization, and initial susceptibility. For example, the zero-field specific heat for

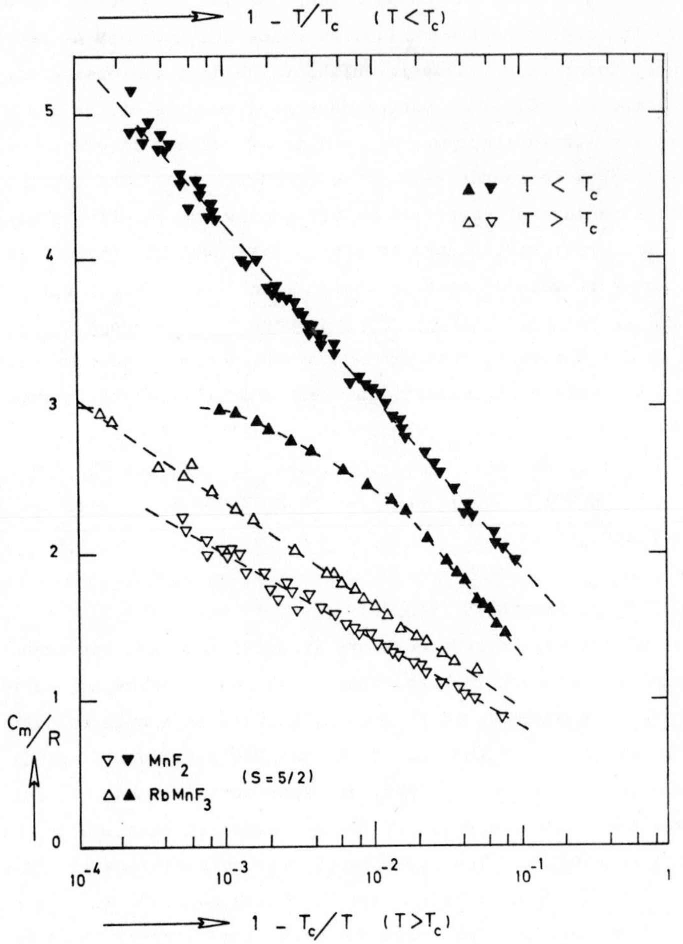

FIGURE 5.13 *The magnetic heat capacity of $RbMnF_3$ and MnF_2 in the neighborhood of T_c (data of Teany). From Ref. 2.*

$T > T_c$ should vary as

$$C/R \approx A(1 - T_c/T)^{-\alpha}$$

where $\alpha = 0$ for the molecular field model, 1/8 for the Ising 3-d model, and about -0.1 for the 3-d Heisenberg model.

G. $Cu(NO_3)_2 \cdot 2\frac{1}{2}H_2O$

We return to this salt to discuss one of the most elegant, recent experiments (21) in magnetism, a field-induced magnetic ordering. Recall, from the discussion in the last chapter, that this compound acts mag-

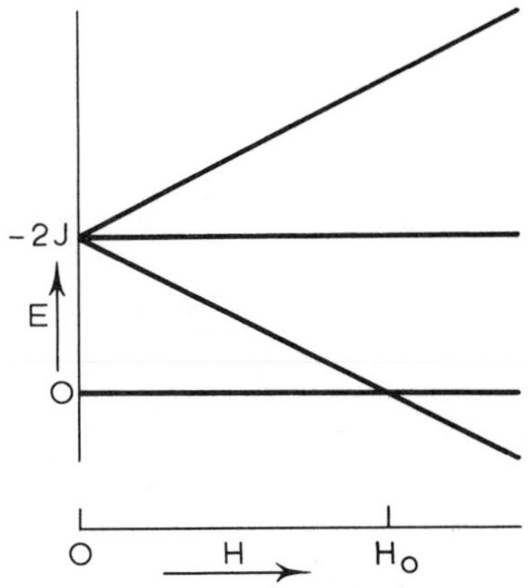

FIGURE 5.14 *Energy levels for an isolated pair of S = 1/2 spins.*

netically as an assemblage of essentially non-interacting dimers. The material cannot undergo long-range ordering under normal conditions because the pair-wise interaction offers a path to remove all the entropy as the temperature goes to zero. But, consider the behavior in a field, under the Hamiltonian

$$\mathcal{H} = -2J\vec{S}_1 \cdot \vec{S}_2 + g\mu_B\vec{H} \cdot (\vec{S}_1 + \vec{S}_2)$$

which may be considered isotropic as long as J is itself isotropic. As illustrated in Figure 5.14, the effect of a field on this system is quite straightforward, and notice in particular the level crossing that occurs as the field reaches the value $H_o = 2|J|/g\mu_B$. At this point, the pairs may be thought of in terms of a system with effective spin S' = 1/2, in an effective field, $H_{eff} = 0$. If this happens, and it does (20), the salt

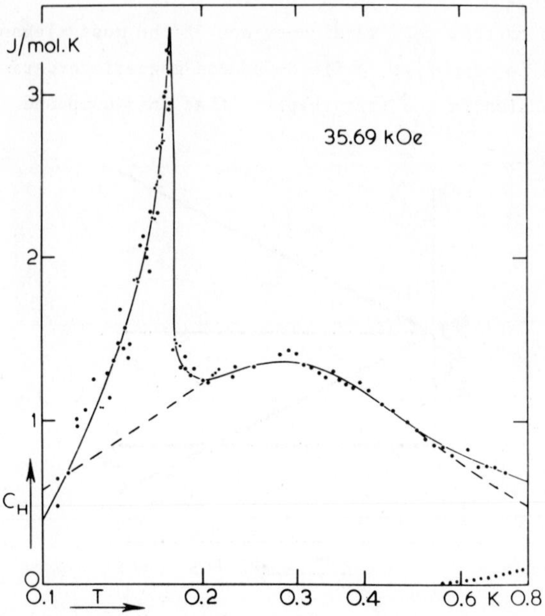

FIGURE 5.15 *Specific heat vs. temperature of $Cu(NO_3)_2 \cdot 2\frac{1}{2}H_2O$ in an external magnetic field of 35.7 kOe. The dashed curve represents the short-range order contribution. The dotted curve is the contribution of the higher triplet levels to the specific heat. From Ref. 21.*

can undergo a phase transition to long-range order by interpair interaction. The value of H_o actually differs from the above value, for it must be corrected for the interpair exchange interaction. The beauty of this system lies with the fact that H_o need only be about 35 kOe, which is easily accessible, in contrast with most other copper dimers known so far, where H_o must be some 2 orders of magnitude larger.

Van Tol (21) used both NMR and heat capacity measurements to observe the transition to long-range order. The specific heat of $Cu(NO_3)_2 \cdot 2\frac{1}{2}H_2O$ in an external field of 35.7 kOe is shown as a function of temperature in Figure 5.15. A λ-like anomaly is observed at 175 mK, indicating the critical temperature; the broad maximum at higher temperatures is due to the short-range order that accumulates in this polymeric material, and is like those effects described in the next chapter. Only about 30 % of the magnetic entropy change is in fact found associated with the long-range ordering.

Since H_o = 35.7 kOe defines H_{eff} as zero, one last point of interest concerns the effect of a non-zero H_{eff}. As one would expect, in either the positive sense (H_o > 35.7 kOe) or negative sense (H_o < 35.7 kOe), the λ-peak shifts to lower temperature and the broad short-range order maximum shifts to higher temperature. That is, the influence of the inter-dimer interaction decreases as H_{eff} becomes non-zero. For H_{eff} > ± 6 kOe, the λ-peak disappears and there is no more long-range order. These results may be thought of as tracing out the phase diagram of the copper compound in the H-T plane. A similar experiment, which has recently been carried out on $[Ni(C_5H_5NO)_6](ClO_4)_2$, will be discussed in Chapt. VIII.

H. DIPOLE-DIPOLE INTERACTIONS

We have so far only considered exchange as the source of interactions between metal ions. While this is undoubtedly the main phenomenon of interest in most cases, magnetic dipole-dipole interactions can be important in some instances, particularly at very low temperatures. Weiss showed long ago that dipole-dipole forces cannot cause ordering at ambient temperatures.

Writing a magnetic moment as

$$\vec{\mu} = g\mu_B \vec{S}$$

then the classical Hamiltonian for dipole-dipole interaction is of the form

$$\mathcal{H}_{d-d} = \sum_{\substack{i,j \\ i \neq j}} [\vec{\mu}_i \cdot \vec{\mu}_j / r_{ij}^3 - 3(\vec{\mu}_i \cdot \vec{r}_{ij})(\vec{\mu}_j \cdot \vec{r}_{ij})/r_{ij}^5] \qquad (5.23)$$

where $\vec{r}_{ij}$ is the vector distance between the magnetic atoms i and j. Since the spin of an ion enters Eq. (5.23) as the square, this effect will be larger, the larger the spin. Notice that solution of this Hamiltonian is an exact, classical type calculation, contrary to what has been expressed above (Eq. (5.22)). Therefore, given the crystal structure of a material, good estimates of the strength of the interaction can be made. Well-known methods (22) allow calculation of the magnetic heat capacity. It is important to observe that λ-like behavior of the specific heat is obtained, irrespective of the atomic source of the ordering. The calculation converges rapidly, for the magnetic specific heat varies as r^{-6}.

The high-temperature tail of the λ anomaly in the heat capacity again takes the form

$$C/R = b/T^2,$$

where the constant b may be calculated from the above Hamiltonian. Accurate measurements (23) on CMN have yielded a value of b of $(5.99 \pm 0.02) \times 10^{-6}\ K^2$, while the theoretical prediction is $(6.6 \pm 0.1) \times 10^{-6}\ K^2$. The close agreement gives strong support to the dipole-dipole origin of the specific heat, although there is available at present no explanation for a heat capacity less than that attributable to dipole-dipole interaction between the magnetic ions. The crystal structure (24) of CMN shows that it consists of $[Mg(H_2O)_6]_3[Ce(NO_3)_6]_2 \cdot 6H_2O$, with the ceriums surrounded by 12 nitrate oxygens at 2.64 Å. These separate the ceriums - the only magnetic ions present - quite nicely, for the closest ceriums are three at 8.56 Å and three more at 8.59 Å. There appears to be no effective superexchange path. More recently T_c was found at 1.9 mK, and the formation of ferromagnetic domains was suggested (25,26).

Magnetic dipole-dipole interaction also appears to be the main contributor to the magnetic properties of such other rare earth compounds as $Er(C_2H_5SO_4)_3 \cdot 9H_2O$, $DyCl_3 \cdot 6H_2O$, $Dy(C_2H_5SO_4)_3 \cdot 9H_2O$ and $ErCl_3 \cdot 6H_2O$ (26), and it also affects the EPR spectra of many metal ion systems. This is because the moment or spin enters the calculation as the square, and many of the lanthanides have larger spin quantum numbers than are found with most iron-series ions. While exchange interaction narrows EPR lines, dipole-dipole interactions broaden them, and this is the reason diamagnetic diluents are usually used for recording EPR spectra. Dipole-dipole interactions set up local fields, typically of the order of 50 or 100 Gauss. With many magnetic neighbors, these fields tend to add randomly at a reference site, and may be considered as a field additional to the external field. Thus, the effective magnetic field varies from site to site, which in turn causes broadening.

Lastly, some anisotropy is often observed in the susceptibilities of manganese(II) compounds. The effect is observed at temperatures too high to be caused by zero-field splittings, and anisotropic exchange effects are not characteristic of this ion. Rather, dipole-dipole interactions are usually assigned (27) as the source of the anisotropy, and this is important because of the high spin (5/2) of this ion.

I. EXCHANGE EFFECTS ON PARAMAGNETIC SUSCEPTIBILITIES

A recurring problem concerns the effect of magnetic exchange on the properties of paramagnetic substances. That is, as the critical temperature is approached from above, short-range order accumulates and begins to influence, for example, the paramagnetic susceptibilities. It is relatively easy to account for this effect in the molecular field approximation (28,29), and this provides a procedure of broad applicability.

Consider a nickel ion, to which Eqs. (3.16-21) would be applied to fit the susceptibilities as influenced only by a zero-field splitting. To include an exchange effect, a molecular exchange field is introduced. This field is given by

$$H_i' = \frac{2zJ}{Ng_i^2 \mu_B^2} \chi_i' H_i ,$$

with $i = \parallel$ or $\perp$ and where χ_i' is the exchange-influenced susceptibility actually measured and where the external field H_i and the resulting exchange field are in the <u>i</u> direction. For convenience, we assume axial symmetry and isotropic molecular g values. Then, with this additional exchange field existing when there is a measuring field, the measured magnetization in the <u>i</u> direction is given by

$$M = \chi_i(H_i + H_i').$$

But then, since by definition the measured susceptibility is given by

$$\chi_i' = \lim_{H_i \to 0} M_i/H_i$$

the exchange-corrected susceptibility is given by

$$\chi_i' = \frac{\chi_i}{1 - (2zJ/Ng_i^2 \mu_B^2)\chi_i}$$

Fits to experimental data thereby allow an evaluation of zJ/k.

More generally, this procedure can be used whenever a theoretical expression for χ_i is available; if the symmetry is lower than axial, the i-directions can refer to any particular set of axes in conjunction with the direction cosines of the molecules with respect to those crystal axes.

J. SUPEREXCHANGE

Some of the ideas concerning superexchange mechanisms were introduced in Chapt. IV-A while, in a sense, this whole book discusses the subject. Direct exchange between neighboring atoms is of course important in metals but this mechanism is generally not of importance with transition metal complexes. The idea that magnetic exchange interaction in transition metal complexes proceeds most efficiently by means of cation-ligand-cation complexes is implicit throughout the discussion, but the problem of calculating theoretically the magnitude of the superexchange interaction for a given cation-ligand-cation configuration to a reasonable accuracy is a difficult task. A variety of empirical rules have been developed, and a discussion of these as well as of the theory of superexchange is beyond the purposes of this book. A recent reference to articles and reviews is available (30).

It is of interest, however, to point out recent progress in one direction, and that is an empirical correlation of exchange constants for a 180° superexchange path M^{2+}-F-M^{2+} in a variety of 3d metal (M = Mn, Co, or Ni) fluorine compounds (30). Two series of compounds were investigated, AMF_3 and A_2MF_4, where A = K, Rb, or Tl, and M = Mn, Co, or Ni. Taking exchange constants from the literature, plots of these were made as a function of metal-ligand separation. For the series of compounds for which data are available, it was found that $|J/k|$ has an R^{-n} dependence, where R is the separation between the metal ions, and n is approximately 12. Thus, exchange interactions are remarkably sensitive to the separation between the metal ion centers.

Of further interest is a correlation of exchange constant with metal ion. By fixing R at a constant value or constant bond length for the series of similar compounds, the ratio of J/k values for Mn^{2+}, Co^{2+} (as spin 3/2) and Ni^{2+} was found as 1:3.6:7.7. While these numbers are only valid for a particular (180°) configuration in fluoride lattices, they provide a useful rule-of-thumb for other situations.

REFERENCES

1. A.H. Morrish, "Physical Principles of Magnetism," John Wiley and Sons, New York, 1965.
2. L.J. de Jongh and A.R. Miedema, Adv. Phys. 23, 1 (1974).
3. H.E. Stanley, "Phase Transitions and Critical Phenomena," Oxford University Press, Oxford, 1971.
4. H.M. Rosenberg, "Low Temperature Solid State Physics," Oxford University Press, Oxford, 1963.
5. D.H. Martin, "Magnetism in Solids," MIT Press, Cambridge, Mass., 1967.

6. A. Danielian, Proc. Phys. Soc. (London) 80, 981 (1962).
7. T.F. Connolly and E.D. Copenhaver, "Bibliography of Magnetic Materials and Tabulation of Magnetic Transition Temperatures," IFI/Plenum, New York, 1972.
8. J.E. Rives, Transition Metal Chem. 7, 1 (1972).
9. C. Domb and A.R. Miedema, Progr. Low Temp. Phys. 4, 296 (1964).
10. M.E. Fisher, Proc. Roy. Soc. (London) A254, 66 (1960); M.E. Fisher, Phil. Mag. 7, 1731 (1962).
11. J. Skalyo, Jr., A.F. Cohen, S.A. Friedberg, and R.B. Griffiths, Phys. Rev. 164, 705 (1967).
12. W.K. Robinson and S.A. Friedberg, Phys. Rev. 117, 402 (1960).
13. A.R. Miedema, R.F. Wielinga and W.J. Huiskamp, Physica 31, 1585 (1965).
14. K. Onô and M. Ohtsuka, J. Phys. Soc. Japan 13, 206 (1958).
15. R.F. Wielinga, H.W.J. Blöte, J.W. Roest, and W.J. Huiskamp, Physica 34, 223 (1967).
16. H. Forstat, J.N. McElearney, and P.T. Bailey, Phys. Lett. 27A, 549 (1968).
17. H.W.J. Blöte and W.J. Huiskamp, Phys. Lett. A29, 304 (1969); H.W.J. Blöte, Thesis, Leiden, 1972.
18. B.N. Figgis, M. Gerloch and R. Mason, Acta Cryst. 17, 506 (1964).
19. W.D. van Amstel, L.J. de Jongh, and M. Matsura, Solid State Comm. 14, 491 (1974).
20. R.F. Wielinga, Prog. Low Temp. Phys., Vol. VI, edited by C.J. Gorter (Amsterdam: North Holland), 1971.
21. M.W. van Tol, K.M. Diederix, and N.J. Poulis, Physica 64, 363 (1973).
22. R.P. Hudson, "Principles and Application of Magnetic Cooling," North-Holland, Amsterdam, 1972.
23. R.P. Hudson and E.R. Pfeiffer, J. Low Temp. Phys. 16, 309 (1974).
24. A. Zalkin, J.D. Forrester and D.H. Templeton, J. Chem. Phys. 39, 2881 (1963).
25. K.W. Mess, J. Lubbers, L. Niesen, and W.J. Huiskamp, Physica 41, 260 (1969).
26. E. Lagendijk, Thesis, Leiden, 1972; E. Lagendijk and W.J. Huiskamp, Physica 65, 118 (1973).
27. F. Keffer, Phys. Rev. 87, 608 (1952); L.R. Walker, R.E. Dietz, K. Andres, and S. Darack, Solid State Comm. 11, 593 (1972); J.N. McElearney, to be published. See also T. Smith and S.A. Friedberg, Phys. Rev. 177, 1012 (1969) for a case where the dipolar term appears not to be the source of the anisotropy.
28. T. Watanabe, J. Phys. Soc. Japan 17, 1856 (1962).
29. J.N. McElearney, D.B. Losee, S. Merchant, and R.L. Carlin, Phys. Rev. B7, 3314 (1973).
30. L.J. de Jongh and R. Block, Physica 79B, 568 (1975).

CHAPTER VI

SHORT-RANGE ORDER

A. INTRODUCTION

Much of what has gone heretofore could come under this classification. Certainly the intracluster magnetic exchange of Chapt. IV falls within this heading, and a significant part of the discussion of Chapt. V also dealt with this subject. "Short-range order" has been defined as "shorter than long-range order," which, while quite accurate, is not very useful. For operational purposes, the term will be restricted to magnetic interactions in one and two dimensions. In that regard, it is interesting to note that the entire discussion is thereby restricted to the paramagnetic region.

The study of magnetic systems which display large amounts of order in but one or two dimensions has been one of the most active areas recently in solid state physics and chemistry (1-6). The reason for this seems to be that, though Ising investigated the theory of ordering in a one-dimensional ferromagnet as long ago as 1925 (7), it was not until recently that it was realized that compounds existed that really displayed this short range order. The first substance recognized as behaving as a linear chain or one-dimensional magnet is apperantly $Cu(NH_3)_4SO_4 \cdot H_2O$ (CTS). Broad maxima at low temperatures were discovered in both the susceptibilities and the specific heat (8). Griffith (9) provided the first quantitative fit of the data, using the calculations for a linear chain of Bonner and Fisher (10). Fortunately, the theoreticians had been busy already for some time (4) in investigating the properties of one-dimensional systems.

B. ONE-DIMENSIONAL OR LINEAR CHAIN SYSTEMS

There is a reasonably good theory available that describes the properties of one-dimensional (1-d) magnetic systems, at least for spin-1/2; what may be more surprising is that there are extensive experimental data as well.

The first point of interest, discovered long ago by Ising (7), is that an infinitely long 1-d system undergoes long-range order only at the temperature of absolute zero. Though Ising used a Hamiltonian that is essentially

$$\mathcal{H} = -2J \sum_i S_{z,i} S_{z,i+1} \tag{6.1}$$

it has been shown subsequently that Heisenberg systems, with the Hamiltonian

$$\mathcal{H} = -2J \sum_i \vec{S}_i \cdot \vec{S}_{i+1} \tag{6.2}$$

likewise do not order at finite temperatures. In reality, of course, interchain actions become more important as the temperature is lowered, and all known 1-d systems ultimately undergo long-range order. But, with $Cu(NH_3)_4SO_4 \cdot H_2O$ as a typical example, T_c = 0.37 K, and so that low temperature makes available at higher temperatures a wide temperature interval where the short-range order effects can be observed.

The specific heat and susceptibilities of an Ising spin-1/2 chain have been calculated exactly in zero-field (11, 12). The calculation results in the following expression for the molar specific heat,

$$C = R\left(\frac{J}{2kT}\right)^2 \operatorname{sech}^2\left(\frac{J}{2kT}\right) \tag{6.3}$$

Note that this function is identical to Eq.(2.14) as plotted in Figure 2.2, if $g\mu_B H$ is replaced by J. Eq.(6.3) is an even function of J, and so the specific heat does not allow one to distinguish ferromagnetic from antiferromagnetic behavior. The compound $CoCl_2 \cdot 2py$, where py is pyridine (C_5H_5N), offers a good illustration of this situation. The structure (13) of this compound is sketched in Figure 6.1, where it will be seen to have a structure of a chain of cobalt atoms bridged by two chlorine atoms. In <u>trans</u>-position are found the two pyridine molecules, forming a distorted octahedral configuration. This basic structure is quite common and will occupy a large part of our discussion of 1-d systems. The specific heat of the compound (1, 14) is illustrated in Figure 6.2; the compound seems to undergo ferromagnetic interaction, and the long-range ordering at 3.15 K complicates the situation, but clearly the breadth of the peak suggests that a large amount of (Ising) short-range order is present.

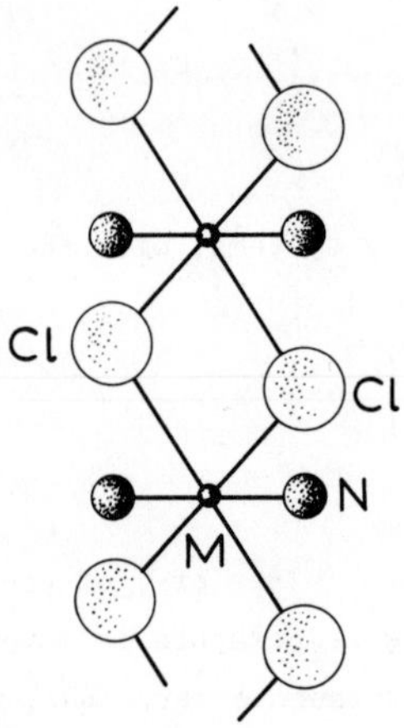

FIGURE 6.1
The -MCl$_2$- chain in the ac plane. From Ref. 14.

The one dimensional effect is illustrated nicely upon comparing in Figure 6.3, on a reduced temperature scale, the magnetic entropies obtained from the specific heats of $CoCl_2 \cdot 2py$ and $CoCl_2 \cdot 2H_2O$. The hydrate is of a structure (15) similar to that of the pyridine adduct, but with water molecules in place of the pyridine molecules. Replacement of water by pyridine would be expected to enhance the magnetic 1-d character, for the larger pyridines should cause an increased interchain separation. Not only does T_c drop from 17.2 K for the hydrate to 3.15 K for the pyridinate, but it will be seen that 60% of the entropy is obtained by $CoCl_2 \cdot 2H_2O$ below T_c, while only 15% of the entropy occurs below T_c for $CoCl_2 \cdot 2py$. The exchange interaction within the chains is essentially the same in the two compounds, and so these effects must be ascribed to a more ideal 1-d (that is, less 3-d) character in the pyridine adduct.

The zero-field susceptibilities have been derived by Fisher (16) for the spin-1/2 Ising chain. They are

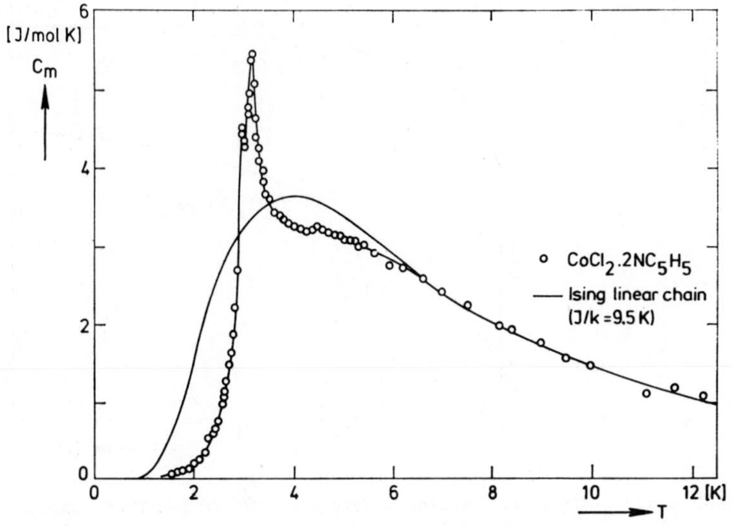

FIGURE 6.2 *Magnetic specific heat versus temperature of $CoCl_2 \cdot 2C_5H_5N$. The curve is the theoretical prediction for the S = 1/2 Ising chain calculated with J/k = 9.5 K. From Ref. 14.*

$$\chi_{\parallel} = \frac{N g_{\parallel}^2 \mu_B^2}{4J} (J/kT) \exp(2J/kT) \tag{6.4}$$

$$\chi_{\perp} = \frac{N g_{\perp}^2 \mu_B^2}{8J} [\tanh(J/kT) + (J/kT) \operatorname{sech}^2(J/kT)] \tag{6.5}$$

It should be noticed that $\chi_{\parallel}$ is an odd function of J, but $\chi_{\perp}$ is an even function. More importantly, the meaning of the symbols "parallel" and "perpendicular" should be emphasized. In the present context, they refer to the external magnetic field direction with respect to the direction of spin-quantization or alignment in the chains, rather than to the chemical or structural arrangement of the chains. In most of the examples studied to date, the spins prefer an arrangement perpendicular to the chain direction. The specific heat, Eq.(6.3) and susceptibilities of Eqs.(6.4) and (6.5) are illustrated for a representative set of parameters in Figure

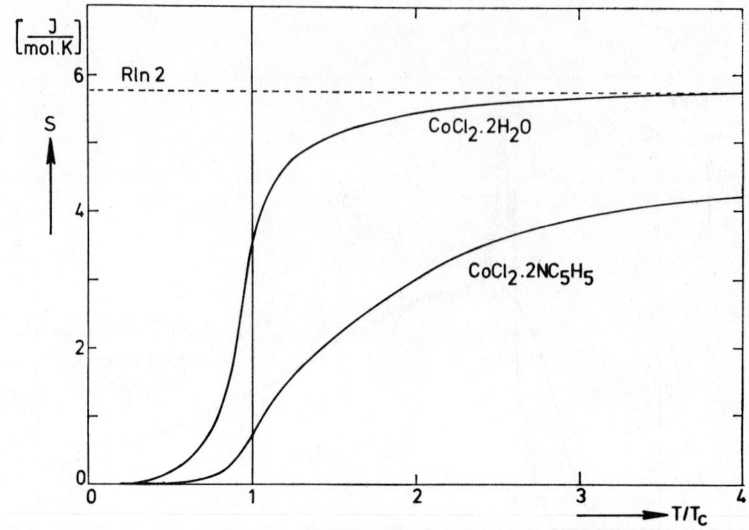

FIGURE 6.3 *Comparison of the entropy versus temperature curves of two Co salts. The large enhancement of the 1-d character by substituting the C_5H_5N molecules for the H_2O molecules can be clearly seen from the large reduction in the amount of entropy gained below T_c. From Ref. 1.*

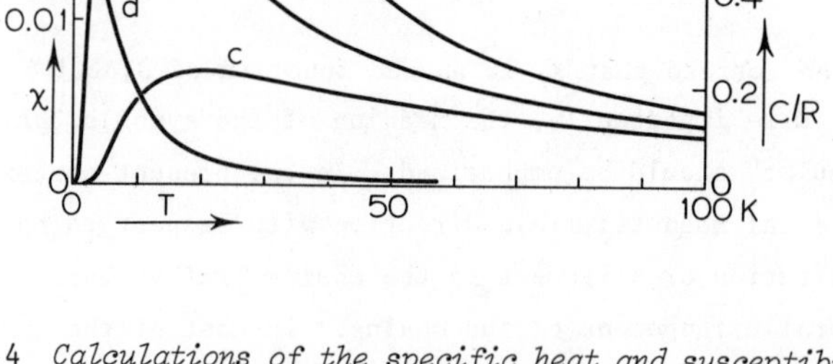

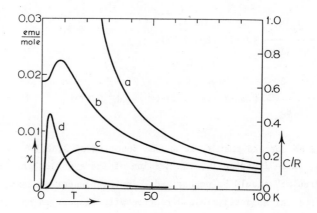

FIGURE 6.4 *Calculations of the specific heat and susceptibilities of Ising linear chains according to Eqs. (6.3) - (6.5). In all cases, $|J/k| = 10$ K; $S = 1/2$, $g_{\parallel} = g_{\perp} = 2$. a : $\chi_{\parallel}(J>0)$; b : $\chi_{\perp}$; c : $\chi_{\parallel}(J<0)$; d : c_p.*

6.4. A characteristic feature of most of the curves is a broad maximum at temperatures comparable to the strength of the exchange interaction; the lone exception is the susceptibility for a measuring field parallel to a ferromagnetic spin alignment.

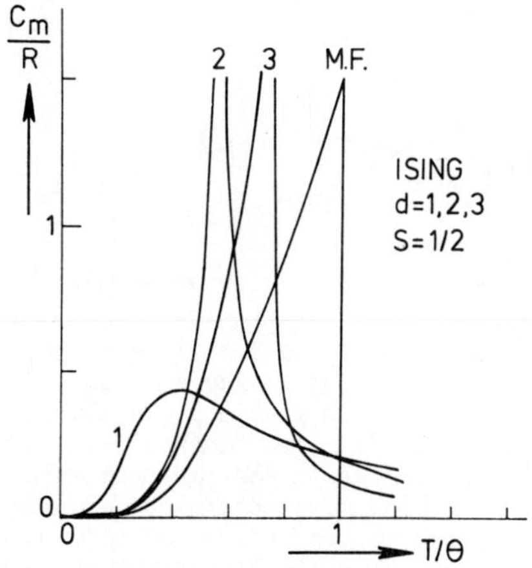

FIGURE 6.5 *Theoretical magnetic specific heats C_m of the $S = 1/2$ Ising model for a 1,2 and 3-d lattice. The chain curve has been obtained by Ising (1925), who first performed calculations on the model that bears his name. The 2-d curve is also an exact result, derived by Onsager (1944) for the quadratic lattice. The 3-d curve has been calculated by Blöte and Huiskamp (1969) and Blöte (1972) for the simple cubic lattice from the high and low-temperature series expansions of C_m given by Baker et al. (1963) and Sykes et al. (1972). For comparison the molecular field prediction (MF) has been included. From Ref. 1.*

Figure 6.5, which is taken again from the review of de Jongh and Miedema (1), illustrates the effect of lattice dimensionality on the $S = 1/2$ Ising model. As was noted earlier, the dimensionality usually exerts a greater influence on the thermodynamic functions than does the variation of lattice structure within a given dimensionality. It will be seen that a decrease in dimensionality increases the importance of short range order effects. The molecular field calculation, which is based on an infinite number of neighbors, takes no account at all of short-range order. In the 1-d case, all of the entropy must be acquired by short range order, for T_c in the ideal case is 0 K. In a simple quadratic (2-d)

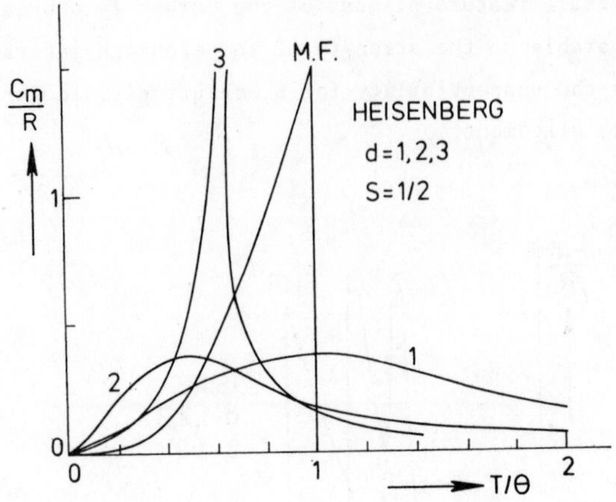

FIGURE 6.6 *Specific heats of the S = 1/2 Heisenberg model in 1,2 and 3 dimensions. The 1-d curve is the result for the antiferromagnetic chain obtained by Bonner and Fisher (1964), from approximate solutions. The 2-d curve applies to the ferromagnetic quadratic lattice and has been constructed by Bloembergen (1971) from the predictions of spin-wave theory (T/θ < 0.1), from the high-temperature series expansion (T/θ > 1), and from the experimental data on approximants of this model (0.1 < T/θ < 1). The 3-d curve follows from series expansions for the b.c.c. ferromagnet given by Baker et al. (1967). Also included is the molecular field prediction. From Ref. 1.*

lattice, 44% of the entropy is acquired below T_c, while 81% of the entropy is acquired below T_C in the case of the 3-d, simple cubic lattice.

There are no exact or closed-form solutions for the Heisenberg model, even for a spin-1/2 one-dimensional system. Nevertheless, machine calculations are available which characterize Heisenberg behavior to a high degree of accuracy, particularly in one-dimension (10, 17). The specific heat of the Heisenberg model in 1, 2, and 3 dimensions is illustrated in Figure 6.6. It will be noticed that not only does the 1-d system not have a non-zero T_c, but even the 2-d Heisenberg system does not undergo long-range order at a non-zero temperature. Comparison of Figures 6.5 and 6.6 will show (1) that changing the type of interaction from the anisotropic Ising to the isotropic Heisenberg form has the effect of enhancing the short-range order contributions. Furthermore, one-dimensional short-range order effects are extended over a much larger region in temperature for the more or less isotropic systems than for the Ising systems.

The best-known example of an antiferromagnetic Heisenberg chain is $[N(CH_3)_4]MnCl_3$, TMMC, the structure (18) of which is illustrated in Figure 6.7. The structure consists of chains of spin-5/2 manganese atoms bridged by three chloride ions; the closest distance between manganese ions in different chains is 9.15 Å. The first report (19) of 1-d behavior in TMMC has been followed by a veritable flood of papers, reviewed in (1) and (2), for the compound is practically an ideal Heisenberg system. The intrachain exchange constant J/k is about −6.7 K, which is expected to cause a broad peak in the magnetic heat capacity at high temper-

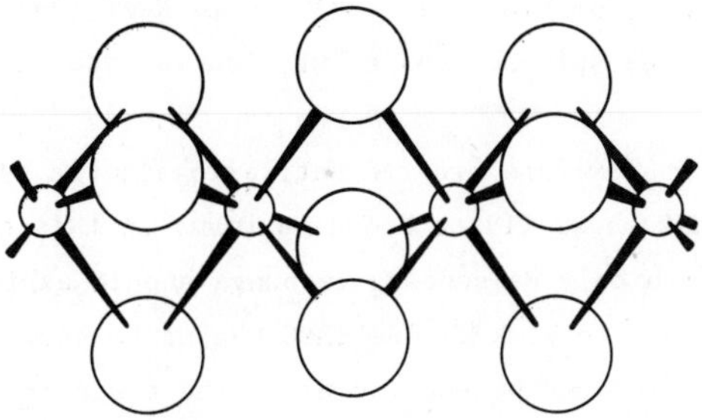

FIGURE 6.7 *A sketch of the linear chains found in* $[N(CH_3)_4]MnCl_3$. *The octahedral environment about the manganese atom is slightly distorted, corresponding to a lengthening along the chain.*

atures. Unfortunately, this relatively large intrachain exchange causes the magnetic contribution to overlap severely with the lattice contribution but, nevertheless, a broad peak with maximum at about 40 K has been separated out (20) by empirical procedures, which depend in part on comparing the measured specific heat of TMMC with its isomorphic, diamagnetic cadmium analog. Long-range order sets in at 0.83 K, as determined both by heat capacity (20-22) and susceptibility (19, 23) measurements. One of the problems faced in analyzing the data is that the theoretical calculations have been hindered by the relatively high value of the spin. One procedure that has been used is based on Fisher's (24) calculation for a classical or infinite spin linear chain, scaled to a real spin of 5/2 (19). This procedure, which was introduced in the first report on linear chain magnetism in $CsMnCl_3 \cdot 2H_2O$ (25), makes use of the equation

$$\chi = \frac{Ng^2\mu_B^2\ S(S + 1)}{3kT}\ \frac{1 - \mu}{1 + \mu} \tag{6.6}$$

where $\mu = (T/T_o) - \coth(T_o/T)$ with $T_o = 2JS(S + 1)/k$. Weng (26) has carried out some calculations that take more explicit cognizance of the spin value, but these remain unpublished. Harrigan and Jones (27) have applied a quantum mechnical correction to Fisher's results that gives a more physically realistic result; for example, the infinite spin model has a nonzero heat capacity at 0 K. Perhaps the most useful calculations are the recent numerical ones of Blöte (17, 28) and de Neef (29), which not only are applicable to the spin-5/2 chain but also include zero-field splitting effects.

As with all short-range ordered antiferromagnetic systems, the susceptibility of TMMC shows (19) a broad maximum, in this case at about 55 K. A system with only Heisenberg exchange should exhibit isotropic susceptibilities, but in weak fields TMMC begins to show anisotropic susceptibilities parallel and perpendicular to the chain (c) axis of the crystal below 60 K. A small dipolar anisotropy was found to be the cause of both the magnetic anisotropy (23) and the alignment of the spins perpendicular to the chain axis.

The compound TMMC is one of a large series of ABX_3 systems, such as $CsCuCl_3$ (30) and $[(CH_3)_4N]NiCl_3$, (TMNC), which display varying degrees of short-range order. Interestingly, TMNC, which is isostructural with TMMC, displays ferromagnetic intrachain interactions, which causes a noncancellation of long-range dipolar fields (31). This result, along with the zero-field splitting of about 3 K, causes TMNC to be a less perfect example of a linear chain magnet than is TMMC. Many of the other physical properties of these systems have recently been reviewed (32).

The XY model (33) is obtained from the Hamiltonian

$$\mathcal{H} = -2J \sum_i (S_{i,x}S_{i+1,x} + S_{i,y}S_{i+1,y}) \tag{6.7}$$

which is again anisotropic; the anisotropy is increased (planar Heisenberg model) by adding a term of the form $D[S_z^2 - \frac{1}{3}S(S + 1)]$ to the above Hamiltonian, for then the spins are constrained to lie in the xy plane. The reader should be careful to distinguish this spin anisotropy from dimensionality anisotropy. There are to date only two examples of one-dimensional compounds that follow this model, $(N_2H_5)_2Co(SO_4)_2$ (34) and Cs_2CoCl_4 (35). Considerable anisotropy in the g-values, with $g_\perp >> g_\parallel$, is the prerequisite for the applicability if the XY model (34,36) and the hydrazin-

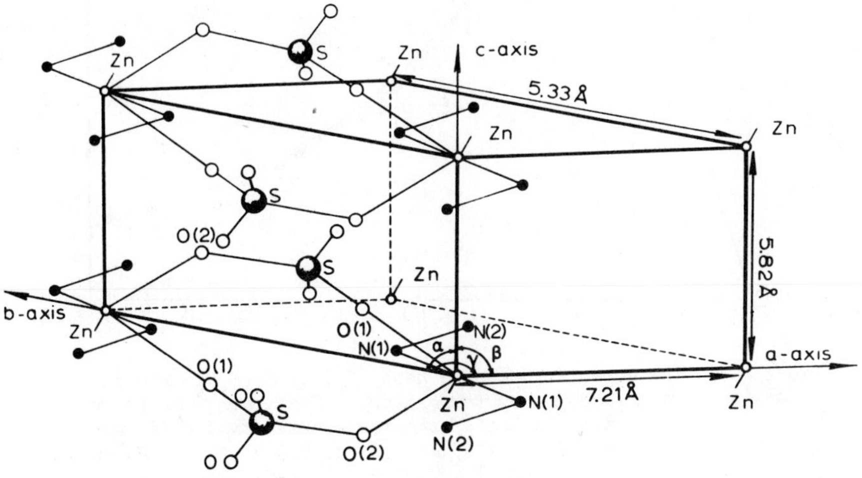

FIGURE 6.8 *Structure of triclinic* $Zn(N_2H_5)_2(SO_4)_2$ *according to Prout and Powell. The chemical linear chain axis (b-axis) is approximately 10 percent shorter than the a-axis. The compounds* $M(N_2H_5)_2(SO_4)_2$ *with M = Ni, Co, Fe and Mn are isomorphous with* $Zn(N_2H_5)_2(SO_4)_2$*. From Ref. 34.*

ium compound, for example meets this restriction, with $g_\perp = 4.9$, $g_\parallel = 2.20$. The compound clearly has a linear chain structure (Fig. 6.8) with metal atoms bridged by two sulfate groups. The magnetic specific heat of $(N_2H_5)_2Co(SO_4)_2$ is illustrated in Figure 6.9, along with a very good fit to the calculated behavior of the XY model linear chain for S = 1/2; the fitting parameter is J/k = -7.05 K. The deviations at lower temperatures are due to the onset of a weak coupling between the chains which causes long-range spin-ordering at 1.60 K.

Several figures taken from de Jongh and Miedema (1) illustrate some of the trends anticipated with 1-d magnets. Figure 6.10 illustrates the heat capacities of a number of chains with S = 1/2, while the perpendicular susceptibilities of the Ising chain and XY models are compared in Figure 6.11. The dependence upon spin-value of the susceptibility and specific heat of the Ising chain is illustrated in Figure 6.12, while this

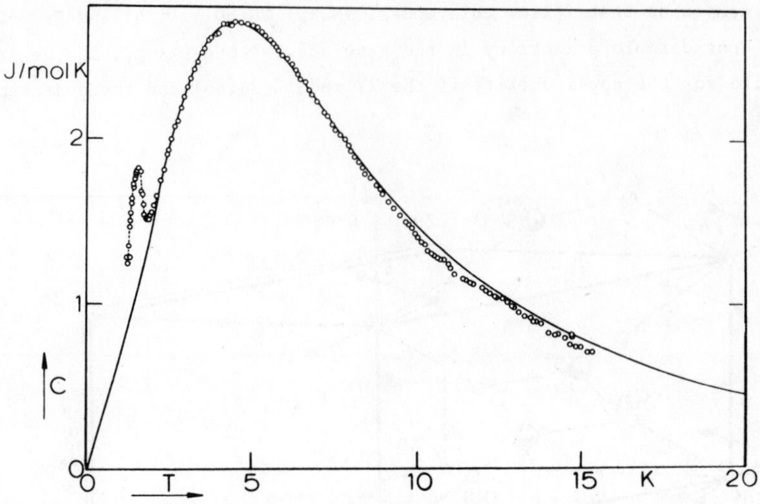

FIGURE 6.9 *The magnetic specific heat, C_m, of $Co(N_2H_5)_2(SO_4)_2$ as a function of temperature. The drawn line represents the results of the transverse coupled linear chain model ($J_\parallel = 0$; $J_\perp = J$) as calculated by Katsura for $S = 1/2$; $|J/k| = 7.05$ K. From Ref. 34.*

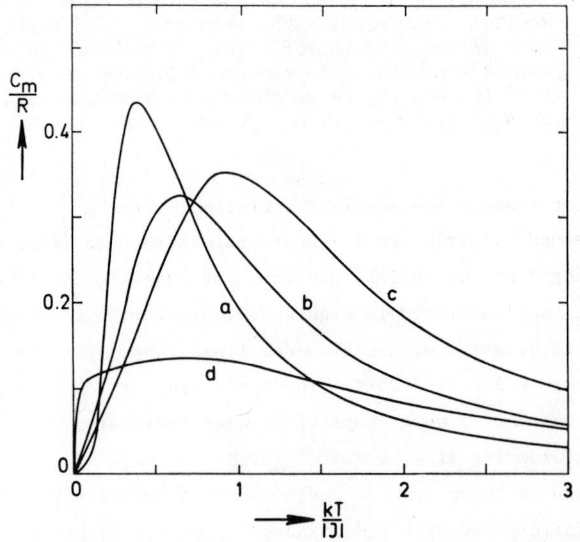

FIGURE 6.10 *Theoretical heat capacities of a number of magnetic chains with $S = 1/2$. Curves (a) and (b) correspond to the Ising and the XY model, respectively (ferro- and antiferromagnetic). Curves (c) and (d) are for the antiferromagnetic and ferromagnetic Heisenberg chain, respectively. From Ref. 1.*

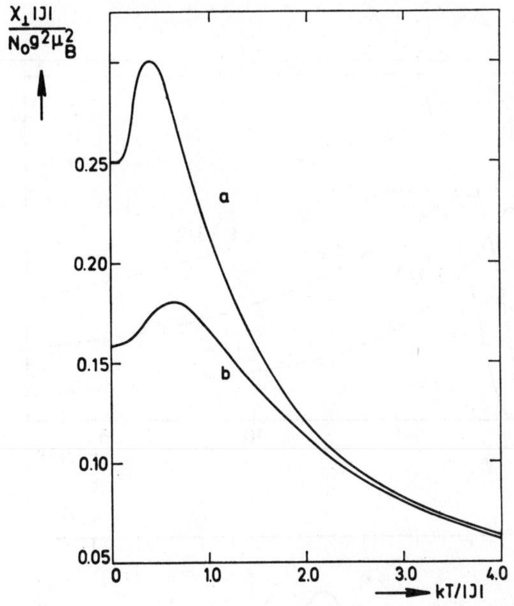

FIGURE 6.11 *Theoretical curves for the perpendicular susceptibility of the S = 1/2 Ising (a) and XY (b) chain model. The values for* $\chi_\perp$ *at* $T = 0$ *are* $Ng^2\mu_B{}^2/2z|J|$ *and* $Ng^2\mu_B{}^2/\pi z|J|$*, respectively. From Ref. 1.*

dependence for the Heisenberg model is shown in Figure 6.13. As it has been pointed out (1), the experimenter is faced with a large number of curves of similar behavior when he seeks to fit experimental data to one of the available models. Recall also (Chapt. IV) that isolated dimers and clusters offer specific heat and susceptibility curves not unlike those illustrated. In fact it has recently been proposed that CTS, the first supposed linear chain magnet, actually behaves more like a two-dimensional magnet. Part of the problem lies with the fact that the physical properties calculated for Heisenberg one- and two-dimensional systems do not differ that much (37). Furthermore, there are no chains or planes of magnetic ions which are self-evident in the structure of CTS.

Clearly, one measurement of a thermodynamic quantity is unlikely to characterize a system satisfactorily. Powder measurements of susceptibility are uniformly unreliable as indicators of magnetic behavior. Crystallographic structures are usually necessary in order to aid in the choice of models to apply, though even here the inferences may not be unambiguous. Thus, $MnCl_2 \cdot 2H_2O$ is a chemical or structural chain but does not exhibit

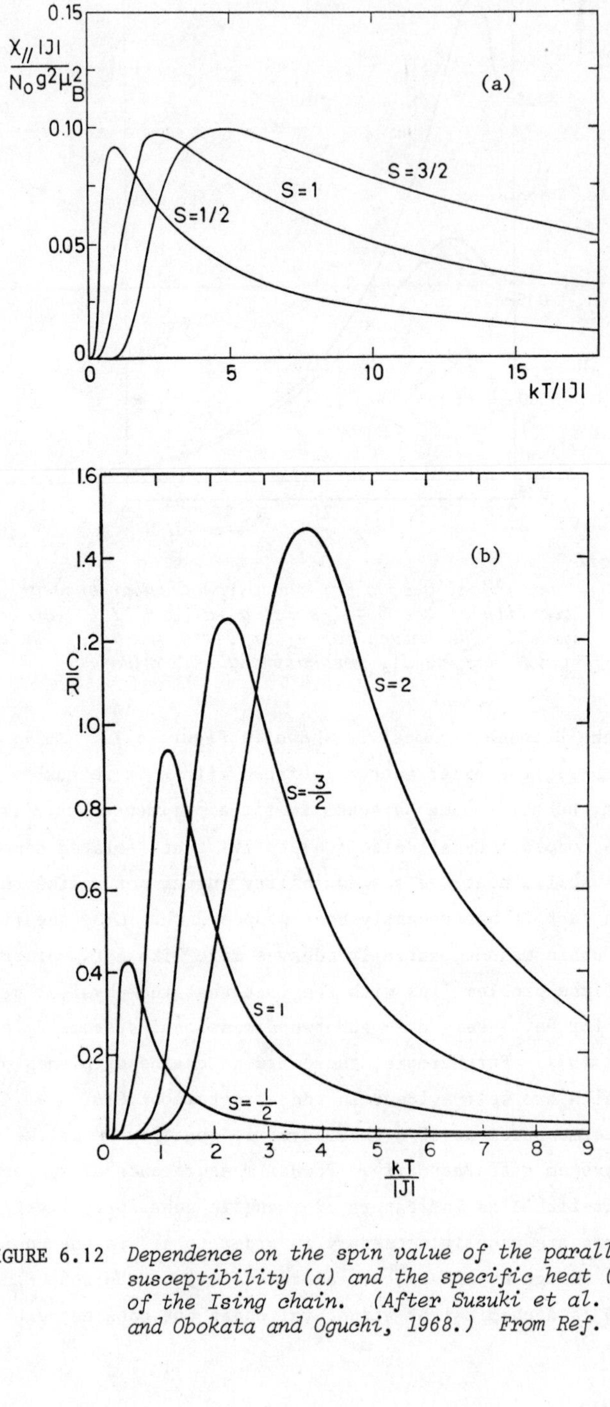

FIGURE 6.12 *Dependence on the spin value of the parallel susceptibility (a) and the specific heat (b) of the Ising chain. (After Suzuki et al. 1967 and Obokata and Oguchi, 1968.) From Ref. 1.*

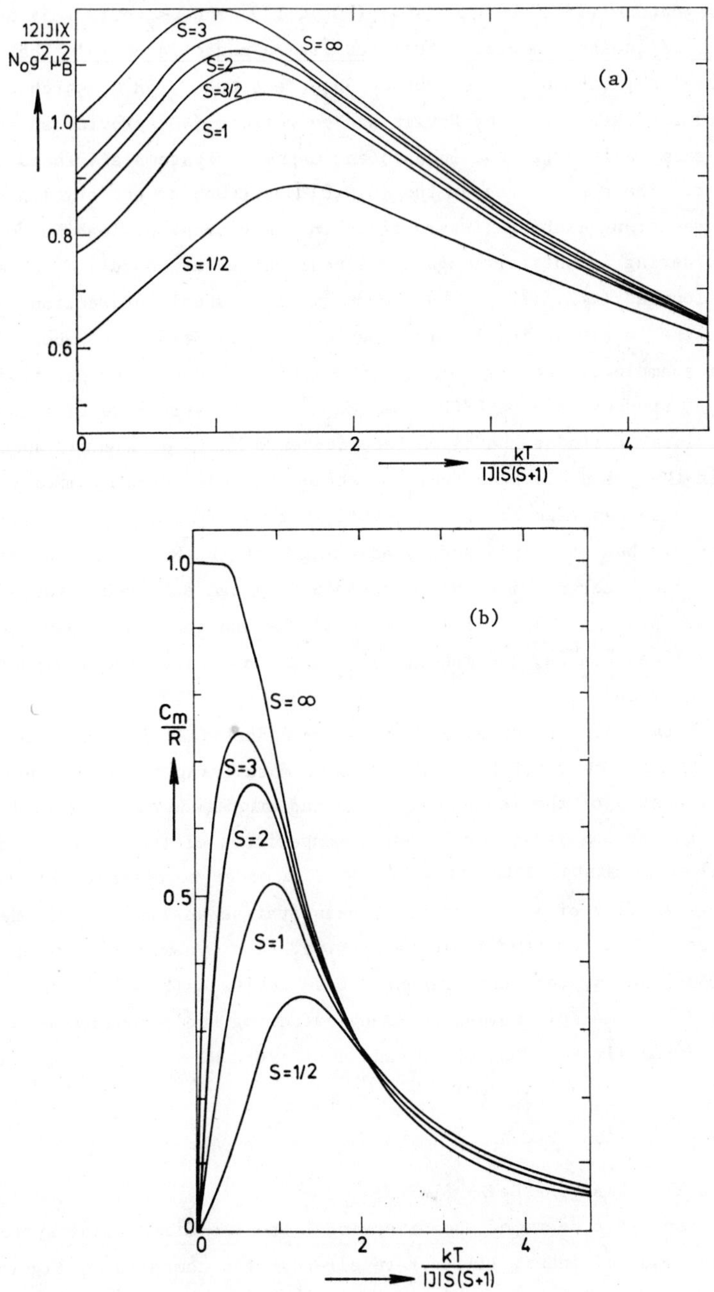

FIGURE 6.13 *Dependence on the spin value of the susceptibility (a) and the specific heat (b) of the Heisenberg chain. (After Weng, 1969.) From Ref. 1.*

chain magnetism (38), while $Cu(NH_3)_4(NO_3)_2$ (39) and Cs_2CoCl_4 (35) each consists of isolated monomers but acts like a magnetic chain! The isostructural series $CuSO_4 \cdot 5H_2O$, $CuSeO_4 \cdot 5H_2O$, and $CuBeF_4 \cdot 5H_2O$, which has been studied extensively by Poulis and co-workers (40) provide an interesting example in which two independent magnetic systems are found to co-exist. The copper ions at the (0,0,0) positions of the triclinic unit cell have strong exchange interactions in one dimension, leading to short-range ordering in antiferromagnetic linear chains at about 1 K. The other copper ions at (1/2, 1/2, 0) have much smaller mutual interactions and behave like an almost ideal paramagnet with Curie-Weiss θ of about 0.05 K. Another example of the sort of complications that can occur is provided by recent studies (41) on $[(CH_3)_3NH]_3Mn_2Cl_7$. The structure of this material consists of linear chains of face-centered $MnCl_6$ octahedra, such as is found in TMMC, and discrete $MnCl_4^{2-}$ tetrahedra; the trimethylammonium cations serve to separate the two magnetic systems. The magnetic susceptibility could best be explained by assuming that the $MnCl_4^{2-}$ ions are formed into linear chains with the intrachain exchange achieved through Cl-Cl contacts. Since the TMMC-like portion of the substance also behaves as a magnetic linear chain, the material in fact consists of two independent linear chain species.

Certain generalizations are useful: manganese(II) is likely to be a Heisenberg ion, cobalt(II) is likely to be more Ising or XY, depending on the geometry of the ion and the sign and magnitude of the zero-field splitting. At the least, qualitative comparisons of the shape of theoretical and experimental data are inadequate in order to characterize the magnetic behavior of a substance. A quantitative analysis of the data and quantitative fit to a model are imperative. The dynamic properties of one-dimensional systems have recently been reviewed (2, 42). The subjects discussed include spin waves, neutron scattering and resonance results, as applied primarily to TMMC, $CsNiF_3$ and $CuCl_2 \cdot 2C_5H_5N$.

C. TWO-DIMENSIONAL OR PLANAR SYSTEMS

The specific heats of one-dimensional and two-dimensional systems in the Ising and Heisenberg limits have already been compared in Figures 6.5 and 6.6, respectively.

The important calculation for two-dimensional systems within the Ising limit is that of Onsager (43) for a spin-1/2 system. Letting J and

J' be the exchange parameters in the two orthogonal directions in the plane, he showed that the quadratic lattice, with J = J', does have a phase transition and exhibits a λ-type anomaly. Furthermore, as soon as any anisotropy is introduced to the one-dimensional system, in which it was pointed out above that no phase transition can occur, a phase transition is now found to occur. That is, with a rectangular lattice even with, say, J/J' = 100, a sharp spike indicative of ordering is found on the heat capacity curve. The specific heats of the linear, rectangular, and quadratic lattice as calculated by Onsager are illustrated in Figure 6.14.

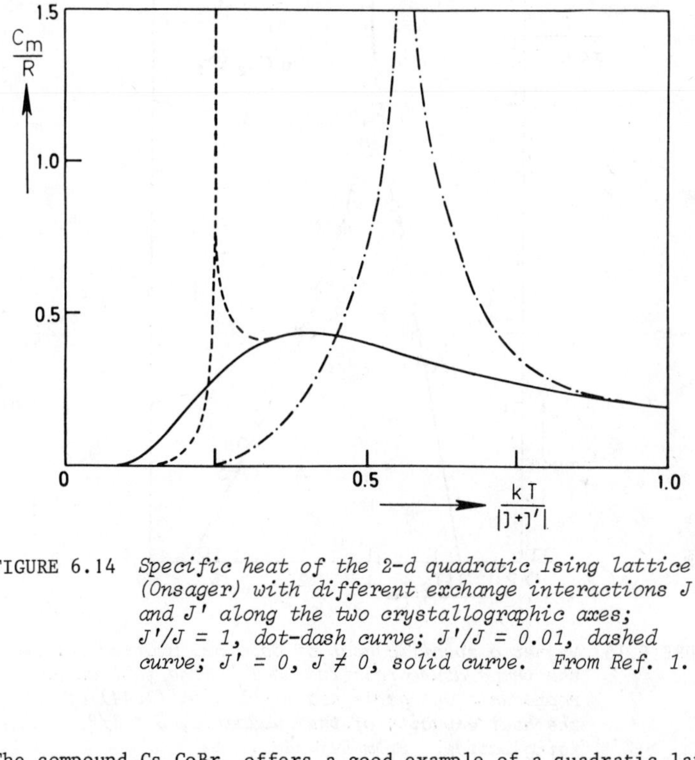

FIGURE 6.14 *Specific heat of the 2-d quadratic Ising lattice (Onsager) with different exchange interactions J and J' along the two crystallographic axes; J'/J = 1, dot-dash curve; J'/J = 0.01, dashed curve; J' = 0, J ≠ 0, solid curve. From Ref. 1.*

The compound Cs_3CoBr_5 offers a good example of a quadratic lattice Ising system (1). This is particularly interesting because the compound is isomorphic to Cs_3CoCl_5, the structure of which was illustrated in Figure 5.11, and which acts as a three-dimensional magnet. Apparently in the bromide there is an accidental cancellation of interactions in the third dimension. The specific heat near T_c (0.282 K) is plotted in Figure 6.15, and compared with Onsager's exact solution for this lattice (44).

There are no exact solutions for the susceptibilities of the planar

Ising lattice, although computer calculations coupled with experimental examples have provided a good approximation for the kind of behavior to be expected. We reproduce in Figure 6.16 a drawing from the article of de Jongh and Miedema that compares the susceptibilities of a quadratic Ising lattice with both molecular field and paramagnetic susceptibilities. Notice that $\chi_\perp$, reflecting the great anisotropy of the Ising model, does not differ from the paramagnetic susceptibility over a large temperature interval. There do not appear to be any calculations of the susceptibility of the rectangular Ising lattice.

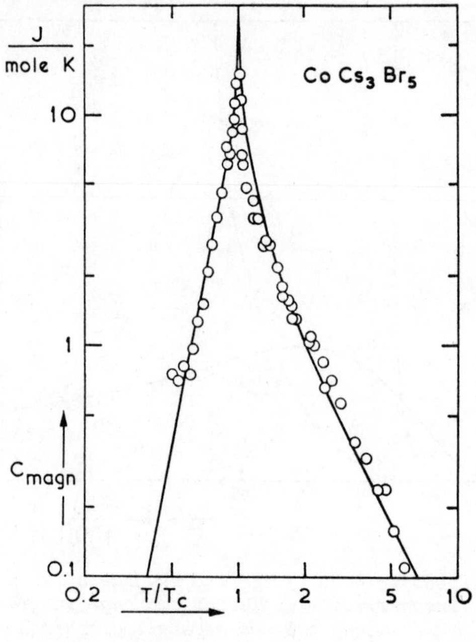

FIGURE 6.15 *Magnetic specific heat of* $CoCs_3Br_5$ *plotted versus the temperature relative to* T_c. *The full curve represents Onsager's exact solution (1944) of the heat capacity of the quadratic,* $S = 1/2$, *Ising latice. From Ref. 44.*

Turning now to isotropic systems, the first point of interest is that it can be proved (1) that a two-dimensional Heisenberg system does not undergo long range order. Thus, as was illustrated above in Figure 6.6, the specific heat of a planar Heisenberg system again consists only of a broad maximum. The shape of the curve differs from that for a one-dimensional system, but nevertheless there is no λ-type anomaly. Of course,

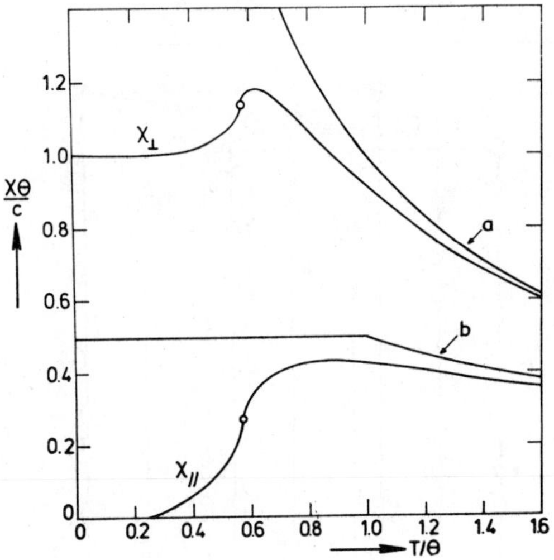

FIGURE 6.16 *A comparison of the perpendicular* ($\chi_\perp$) *and the parallel* ($\chi_\parallel$) *susceptibility of the quadratic Ising lattice (Fisher 1963, Sykes and Fisher 1962) with S = 1/2. Curve a is the susceptibility of a paramagnetic substance. Curve b is the molecular field prediction for the antiferromagnetic susceptibility in the paramagnetic region and for the perpendicular part below the transition temperature. From Ref. 1.*

just as with the experimental examples of one-dimensional systems, nonideal behavior is observed as the temperature of a two-dimensional magnet is steadily decreased. Anisotropy can arise from a variety of sources, and interlayer exchange, though weak, can also eventually become important enough at some low temperature to cause ordering. In many cases, however, the λ-anomaly appears as only a small spike, superimposed on the two-dimensional heat capacity, and the system still acts largely as a two-dimensional lattice even to temperatures well below T_c. Unfortunately, most of the planar magnets investigated to date exhibit their specific heat maxima at high temperatures, and the separation of the lattice contribution has been difficult.

There has been a large amount of success in finding systems that are structurally, and therefore magnetically, two-dimensional. One system investigated extensively at Amsterdam (45) has the stoichiometry $(C_nH_{2n+1}NH_3)_2CuCl_4$, with n having the values 1,2,... all the way to 10. The

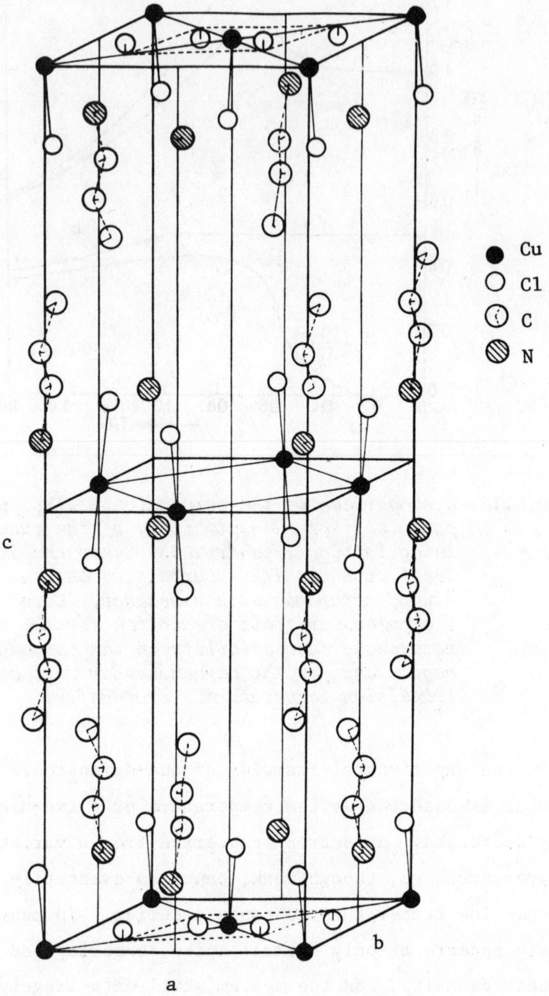

FIGURE 6.17 *Crystal structure of $(C_3H_7NH_3)_2CuCl_4$. Part of the propyl ammonium groups and H atoms have been omitted for clarity. From Ref. 1.*

crystals consist of ferromagnetic layers of copper ions, separated by two layers of non-magnetic alkyl ammonium groups. The representative structure is illustrated in Figure 6.17. By varying n, the distance between copper ions in neighboring layers may be increased from 9.97 Å in the methyl compound to 25.8 Å when n = 10, while the configuration within the copper layers is not appreciably changed. The Cu-Cu distance within the layers is about 5.25 Å. Thus, the interlayer separation can be varied almost at will, while the intralayer separations remain more or less the same. The

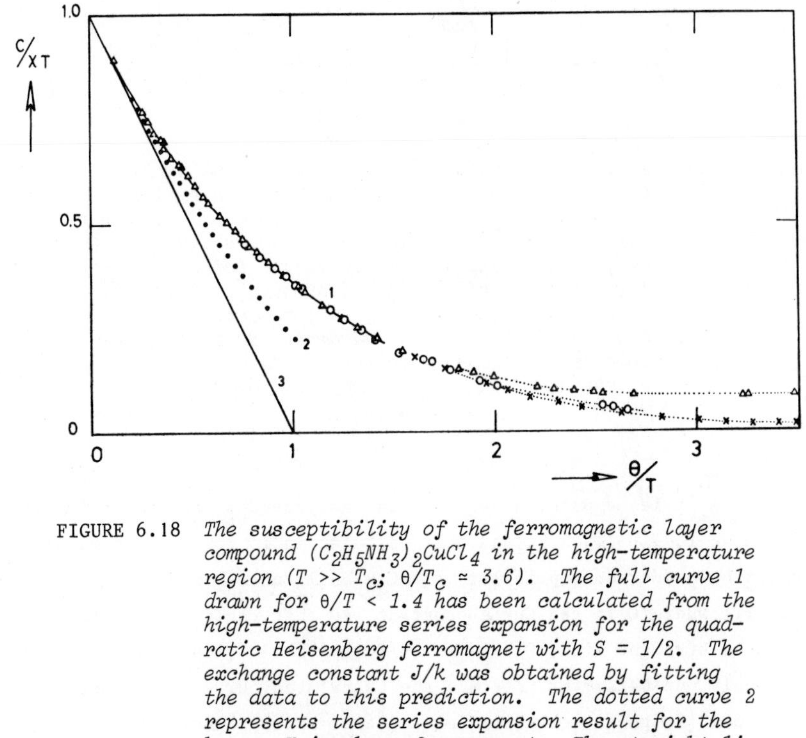

FIGURE 6.18 *The susceptibility of the ferromagnetic layer compound* $(C_2H_5NH_3)_2CuCl_4$ *in the high-temperature region* ($T \gg T_c$; $\theta/T_c \simeq 3.6$). *The full curve 1 drawn for* $\theta/T < 1.4$ *has been calculated from the high-temperature series expansion for the quadratic Heisenberg ferromagnet with* $S = 1/2$. *The exchange constant* J/k *was obtained by fitting the data to this prediction. The dotted curve 2 represents the series expansion result for the b.c.c. Heisenberg ferromagnet. The straight line 3 is the MF prediction* $C/\chi T = 1 - \theta/T$ *for the quadratic ferromagnet. (After De Jongh et al. 1972).*
Δ:*H = 10 kOe. O:H = 4 kOe:* $\times$:*H = 0 (a.c. susceptibility measurements). From Ref. 1.*

manganese analogues are isostructural (1), which allows this to be one of the best understood two-dimensional structures.

The high-temperature susceptibilities of the copper compounds yield positive Curie-Weiss constants, suggesting a ferromagnetic interaction in

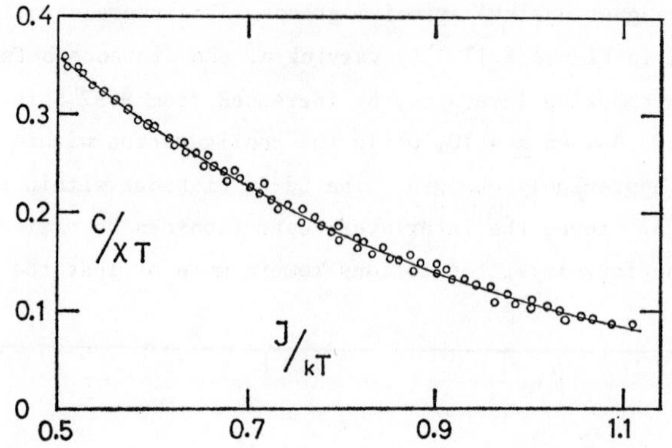

FIGURE 6.19 *Susceptibility data of eleven different members of the series* $(C_nH_{2n+1}NH_3)_2CuX_4$ *in the region* $0.5 < J/kT < 1.1$ $(J/k = \theta/2)$. *Since the results for the various compounds coincide, although they differ in strength and type of the interlayer interaction as well as in anisotropy, the straight line through the data may be considered to represent the* χ *of the ideal quadratic Heisenberg ferromagnet. As such this result may be seen as an extension of the series expansion prediction, which is trustworthy up to* $J/kT < 0.6$ *only. From Ref. 1.*

the plane. A series expansion calculation has provided the susceptibility behavior down to $T \approx 1.5$ J/k, and this calculation is compared in Figure 6.18 with the powder susceptibility of $(C_2H_5NH_3)_2CuCl_4$. Here $C/\chi T$ is plotted vs. θ/T, where C is the Curie constant, so that the Curie-Weiss law appears as a straight line, $C/\chi T = 1 - \theta/T$ (Curve 3). The curve 1, which requires only one parameter to scale the calculated curve to the experimental points, is the calculation for a planar spin-1/2 Heisenberg system; the parameter mentioned is the exchange constant, positive in sign, $J/k = 18.6$ K. The dotted line, curve 2, represents the series expansion result for the (3-d) b.c.c. Heisenberg ferromagnet for comparison. Figure 6.19 presents susceptibility data for 11 compounds of the series $(C_nH_{2n+1}NH_3)_2$-CuX_4. Since the results for the various compounds coincide, although they differ in strength and type of interplane interaction (J') and anisotropy, the curve through the data may be considered to represent the susceptibility of the ideal quadratic Heisenberg ferromagnet, at least in the region $0.5 < J/kT < 1.1$, where $J/k = \theta/2$. Though the intraplane exchange is ferromagnetic for all the salts, J' is antiferromagnetic except for the cases of $\underline{n}$ = 1 and 10. For the ethyl derivative $|J'/J|$ is about 8.5×10^{-4}, which illustrates how weak the interplanar interaction is, with respect to the intraplanar one; furthermore, T_c = 10.20 K.

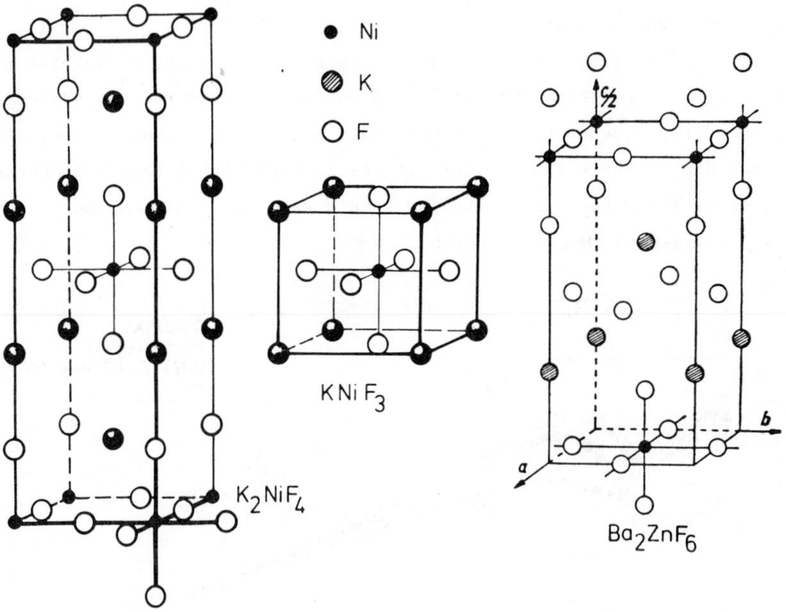

FIGURE 6.20 *Comparison of three related crystal structures, two of which are 2-d in magnetic respect. In the middle the cubic perovskite structure of* $KNiF_3$, *on the left the tetragonal* K_2NiF_4 *unit cell. On the right the structure of* Ba_2ZnF_4 *(Von Schnering 1967). These crystal structures offer the possibility of comparing the 2-d and 3-d properties of compounds which are quite similar in other respects. From Ref. 1.*

Another series of two-dimensional systems that has been widely studied (1) is based on the K_2NiF_4 structure. As illustrated in Figure 6.20, octahedral coordination of the nickel ion is obtained in the cubic perovskite, $KNiF_3$. The tetragonal K_2NiF_4 structure can be looked upon as being derived from the $KNiF_3$ lattice by the addition of an extra layer of KF between the NiF_2 sheets. By this simple fact a 3-d antiferromagnetic lattice ($KNiF_3$) is transformed into a magnetic layer structure (K_2NiF_4). It is of importance that the interaction within the layer is antiferromagnetic, since this causes a cancellation of the interaction between neighboring layers in the ordered state. Similar compounds have been examined with the nickel ion replaced by manganese, cobalt or iron, the potassium by rubidium or cesium, and in some cases fluoride by chloride. The correlation between crystallographic structure and magnetic exchange which the chemist would like to observe is in fact rather remarkably displayed by compounds such as these.

The antiferromagnetic susceptibility of planar magnets exhibits a behavior similar to that of isotropic AF chains, because the two systems have in common the absence of long range order and of anisotropy. Then, a broad maximum due to short range order effects is expected at higher temperatures, whereas at T = 0 the susceptibility should attain a finite value. There is no closed-form theory available.

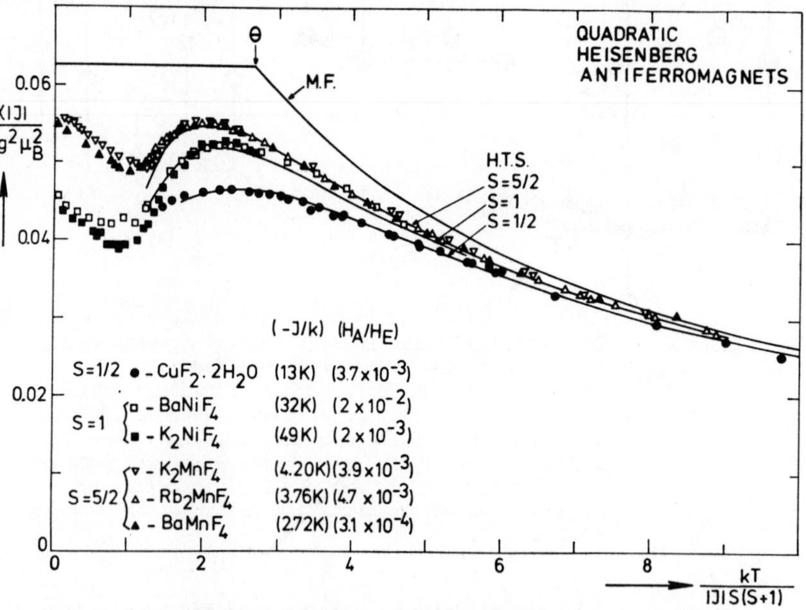

FIGURE 6.21 *The susceptibility of six examples of the quadratic Heisenberg antiferromagnet. The experimental data in the high-temperature region ($T > T_c$: $kT \approx S(S+1)|J|$) have been fitted to the theoretical (solid) curves by varying the exchange constants J/k. These curves have been calculated from the high-temperature series expansions for the susceptibility (H.T.S.). Note the large deviation from the molecular field result (MF) for the susceptibility in the paramagnetic region and below the transition temperature ($\chi_\perp$). Below T_c the measured perpendicular susceptibilities of two S = 5/2 and two S = 1 compounds have been included. The differences between $\chi_\perp$ for compounds with the same S reflect the difference in anisotropy. From Ref. 1.*

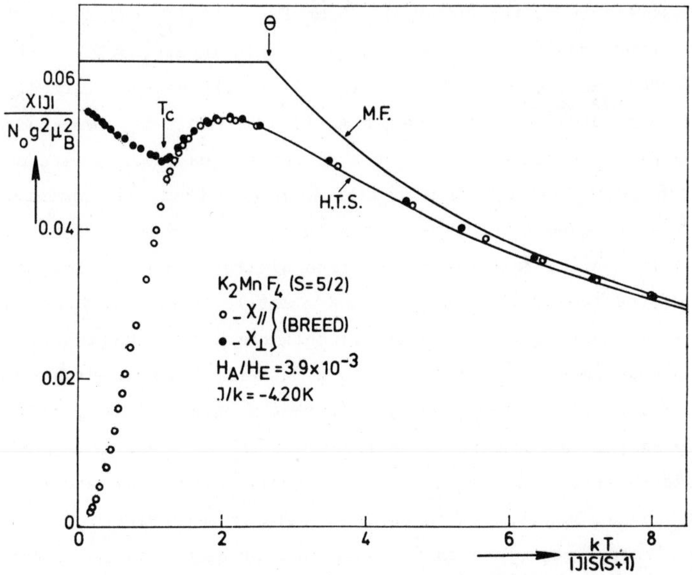

FIGURE 6.22 *The measured parallel and perpendicular susceptibility of K_2MnF_4, which is an example of the quadratic S = 5/2 Heisenberg antiferromagnet. The value of J/k has been determined by fitting the high-temperature susceptibility to the series expansion prediction (H.T.S.). The value of the anisotropy parameter H_A/H_E and the transition temperature T_c have been indicated. From Ref. 46.*

The susceptibilities of six samples that approximate the quadratic antiferromagnetic Heisenberg layer, with different spin values, are illustrated in Figure 6.21; note the large deviation from the molecular field (MF) result for the susceptibility in the paramagnetic region and below the transition temperature, where $\chi_{\perp}$ has been plotted. Note also that the curves deviate more from the MF result, the lower the value of the spin, and therefore the more important are quantum effects. In Figure 6.22, the susceptibilities of K_2MnF_4 are illustrated over a wide temperature range (46). At the transition temperature, T_c = 42.3 K, the parallel and perpendicular susceptibilities will be seen to diverge; the exchange constant, which is obtained by fitting the data to a series expansion prediction, is J/k = -4.20 K. The anisotropy in the plane was estimated as about 3.9×10^{-3}, while the interplanar exchange is small, $|J'/J| \approx 10^{-6}$.

The wide variety of potential magnetic phenomena is illustrated by recent measurements on the compounds $Rb_3Mn_2F_7$, $K_3Mn_2F_7$, and $Rb_3Mn_2Cl_7$ (47, 48). The structure is one of essentially double layers, or thin films: sheets of $AMnX_3$ unit cells, of one unit cell thickness, are formed, separated from each other by non-magnetic AX layers. That is, two quadratic layers are placed upon each other at a distance equal to the lattice spacing within each layer, and are in turn well-separated from the adjoining layers. The consequence of this is that a broad maximum is found in the susceptibility while, furthermore, neutron diffraction shows the behavior expected of the two-dimensional Heisenberg antiferromagnet. Each manganese ion is antiparallel to each of its neighbors. Interestingly, both theory and experiment show that in a thin magnetic film the high-temperature properties are not appreciably different from bulk (i.e., 3-d) behavior. Only as the temperature is lowered does the effect of the finite film thickness become apparent. Indeed, theory (high-temperature series) shows that the susceptibility for a Heisenberg film consisting of four layers is already very close to the bulk susceptibility in the temperature region studied.

The important role that hydrogen-bonding plays in determining lattice dimensionality should not be ignored, and illustrates the importance of a careful examination of the crystal structure. Though crystallographic chains are present in the compounds $[(CH_3)_3NH]MX_3 \cdot 2H_2O$, hydrogen-bonding between the chains causes a measurable planar magnetic character for these molecules, which are described more fully in Chapt. VIII-C-10. Two other compounds investigated recently which obtain planar character largely through hydrogen-bonding that links metal polyhedra together are the copper complex of L-isoleucine (49) and $Rb_2NiCl_4 . 2H_2O$ (50). In the case of Cu-(L-i-leucine)$_2$.$2H_2O$, the material consists of discrete five-coordinate molecules of this stoichiometry, but they are hydrogen-bonded together in two crystalline directions, while well-isolated in the third. The material acts as a two-dimensional Heisenberg ferromagnet. Appreciable short range order was not anticipated in the investigation of $Rb_2NiCl_4 \cdot 2H_2O$, for earlier investigations of the manganese isomorphs had not revealed significant lower-dimensional behavior. Yet, in fact, this material, which consists of the discrete octahedra which usually lead to normal 3-d ordering, exhibits important low-dimensional character. A careful examination of the crystal structure shows that the octahedra are linked by hydrogen-bonds into sheets, which in turn lead to the observed magnetic behavior. It is not yet clear why the manganese compounds behave in a different fashion.

The only apparent examples known to date of planar or 2-d magnets that follow the XY model are $CoCl_2 \cdot 6H_2O$ and $CoBr_2 \cdot 6H_2O$ (36, 51, 52). Again theory predicts that long-range order will set in for this system only at T = 0, but non-ideality in the form of in-plane anisotropy (J_x, J_y and J_z) and interplanar interactions (J') causes three-dimensional ordering to occur.

The structure of both these crystals consists of discrete trans-[$Co(H_2O)_4X_2$] octahedra linked by hydrogen-bonds in a face-centered arrangement as illustrated in Figure 6.23; the two dimensional character arises in the ab (XY) plane, for the molecules are closer together in this direction and provide more favorable superexchange paths than along the c-axis. The g-values are approximately $g_\perp = 5.1$ and $g_\parallel = 2.5$ for the chloride, so there is a strong preference for the moments to lie within a plane that nearly coincides with the $Co(H_2O)_4$ plane. This is a precondition, of course, for the applicability of the XY model. Furthermore, the large de-

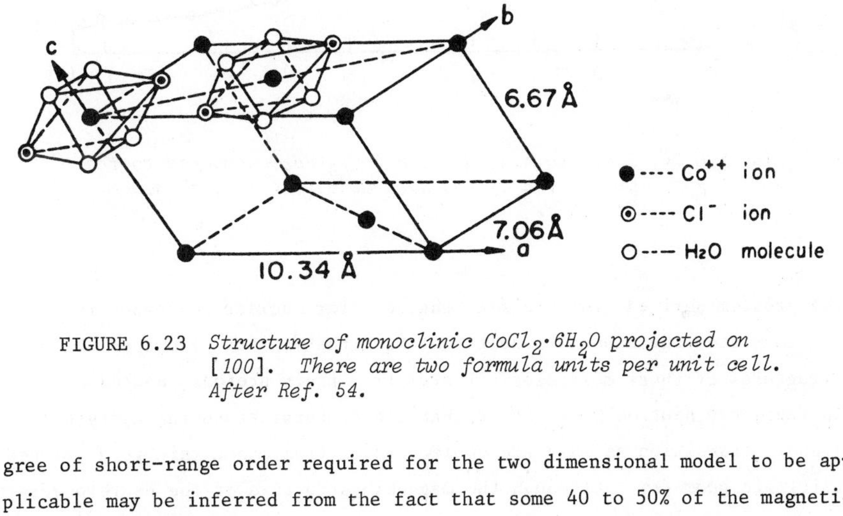

FIGURE 6.23 *Structure of monoclinic $CoCl_2 \cdot 6H_2O$ projected on [100]. There are two formula units per unit cell. After Ref. 54.*

gree of short-range order required for the two dimensional model to be applicable may be inferred from the fact that some 40 to 50% of the magnetic entropy is gained above T_c.

Thus, the specific heat of $CoCl_2 \cdot 6H_2O$ (Fig. 6.24) at high temperatures follows the theoretical curve for the XY model, but a λ-type peak is superimposed on the broad maximum. The parameters are (36) $T_c = 2.29$ K and $J_x/k = -2.05$ K, with only about a 4% anisotropy in the plane, while it is found that $J_z/J_x = 0.35$, which illustrates the substantial intraplane anisotropy. Furthermore $|J'/J|$, where J' is the interplane exchange, is about 10^{-2}.

The magnetic structures of $CoCl_2 \cdot 6D_2O$ and $CoBr_2 \cdot 6D_2O$ have been investigated recently by neutron diffraction (53). Part of the interest in

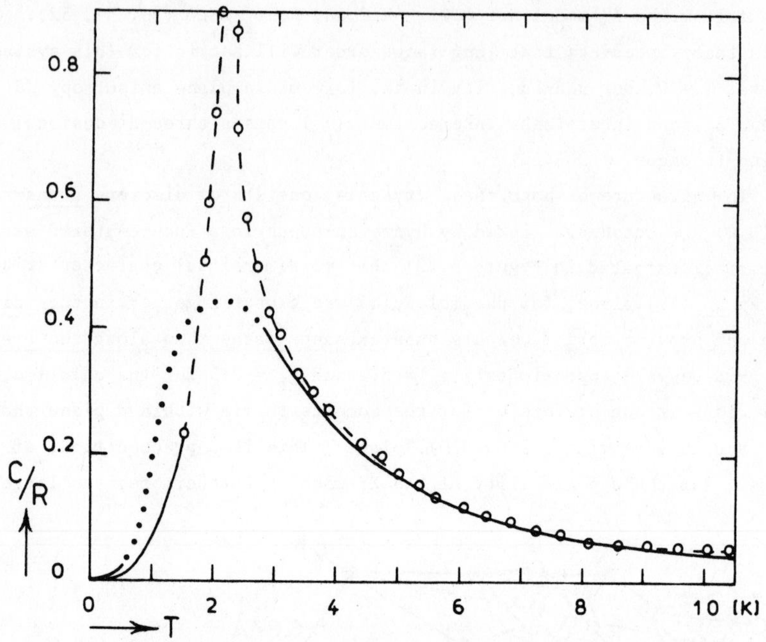

FIGURE 6.24 *Specific heat data on $CoCl_2 \cdot 6H_2O$ compared to the prediction of the quadratic, S = 1/2, XY model. From Ref. 36.*

the problem derived from the fact that earlier nuclear resonance results had suggested the unusual result that both the magnetic and crystallographic structures of these molecules differed from their protonic analogs. In fact, the neutron work showed that the deuterates undergo a crystallographic transition as they are cooled, going from monoclinic to a twinned triclinic geometry. Although the magnetic structure of the deuterohydrates differs slightly from that of the hydrates by a rotation of the moments away from the ac plane over an angle of 45°, both compounds remain as examples of the 2-d XY model.

D. LONG RANGE ORDER

As has been implied above, all real systems with large amounts of short range order eventually undergo long-range ordering at some (low) temperature. This is because ultimately, no matter how small the deviation from ideality may be, eventually it becomes important enough to cause a three-dimensional phase transition. Naturally, any interaction between chains (or planes), no matter how weak, will still cause such a transition. Furthermore, any anisotropy will eventually cause long range ordering. The latter can be due to a small zero-field splitting in manganese(II), for example, which at some temperature becomes comparable to kT, but even effects such as anisotropic thermal expansion are sufficient to cause the transition to long-range order. The sensitivity of (three-dimensional) transition temperatures to single-ion anisotropy has recently been explored theoretically (55).

Nevertheless, the nature of the ordered state differs from the usual 3-d example when it is derived from a set of formerly independent antiferromagnetic linear chains. This is particularly so in the case of S = 1/2 isotropic Heisenberg chains (40), as a result of the extensive spin-reduction ΔS that occurs in these systems. Spin-reduction is a consequence of the low-lying excitations of an antiferromagnetic system which are called spin waves (1). Though we shall not explore these phenomena any further, the fully antiparallel alignment of the magnetic moments is an eigenstate of the antiferromagnetic exchange Hamiltonian only in the Ising limit. An effect of zero-point motions or spin waves is that at T = 0 the magnetic moment per site is no longer $g\mu_B S$, but $g\mu_B(S - \Delta S)$. The spin-reduction $\Delta S(T,H)$ depends on the anisotropy in the exchange interaction and reduces to zero in the Ising limit. The spin-reduction depends inversely on the number of interacting neighbors z and on the spin value S. So, although ΔS is small for 3-d systems, it increases for 2-d systems, and is maximal for linear-chain (z=2), S = $\frac{1}{2}$ Heisenberg antiferromagnets.

Finally, it was pointed out in Chapt. V-D that for antiferromagnets the ordering temperature, T_c, lies below the temperature at which the susceptibility has its maximum value. In fact, for 3-d ordering, the maximum is expected to lie some 5% above T_c. One of the most noticeable characteristics of lowerdimensional ordering is that the broad maximum in the susceptibility occurs at a temperature substantially higher than T_c, as much as 40% for a two-dimensional lattice (56).

REFERENCES

1. L.J. de Jongh and A.R. Miedema, Adv. Phys. 23, 1 (1974).
2. D.W. Hone and P.M. Richards, An.Rev. Material Sci. 4, 337 (1974).
3. R.L. Carlin, Accts. Chem. Res. 9, 67 (1976).

4. E.H. Lieb and D.C. Mattis, "Mathematical Physics in One Dimension," Academic Press, N.Y., 1966.
5. M.E. Fisher, "Essays in "Physics," Ed. G.K.T. Conn and G.N. Fowler, Academic Press, N.Y., Vol. 4, p. 43 (1972).
6. A variety of papers on this subject are included in the symposium volume, "Extended Interactions between Metal Ions in Transition Metal Complexes," ed. L.V. Interrante, American Chemical Society, Symposium Series 5, Washington, D.C., 1974.
7. E. Ising, Zeit. f. Physik 31, 253 (1925).
8. J.J. Fritz and H.L. Pinch, J. Am. Chem. Soc. 79, 3644 (1957); J.C. Eisenstein, J. Chem. Phys. 28, 323 (1958); T. Watanabe and T. Haseda, J. Chem. Phys. 29, 1429 (1958); T. Haseda and A.R. Miedema, Physica 27, 1102 (1961).
9. R.B. Griffiths, Phys. Rev. 135, A659 (1964).
10. J.C. Bonner and M.E. Fisher, Phys. Rev. 135, A640 (1964).
11. G.F. Newell and E.W. Montroll, Rev. Mod. Phys. 25, 353 (1953).
12. B.M. McCoy and T.T. Wu, "The Two-Dimensional Ising Model," Harvard University Press, Cambridge, Mass., 1973.
13. J.D. Dunitz, Acta Cryst. 10, 307 (1957); P.J. Clarke and H.J. Milledge Acta Cryst. B31, 1543 (1975).
14. K. Takeda, S. Matsukawa, and T. Haseda, J. Phys. Soc. Japan 30, 1330 (1971); see also W.J.M. de Jonge, Q.A.G. van Vlimmeren, J.P.A.M. Hijmans, C.H.W. Swüste, J.A.H.M. Buys and G.J.M. van Workum, Phys. Rev., to be published.
15. B. Morosin and E.J. Graeber, Acta Cryst. 16, 1176 (1963).
16. M.E. Fisher, J. Math. Phys. 4, 124 (1963).
17. H.W.J. Blöte, Physica 78, 302 (1974); T. de Neef, A.J.M. Kuipers and K. Kopinga, J. Phys. A 7, L171 (1974).
18. B. Morosin and E.J. Graeber, Acta Cryst. 23, 766 (1967).
19. R. Dingle, M.E. Lines, and S.L. Holt, Phys. Rev. 187, 643 (1969).
20. R.E. Dietz, L.R. Walker, F.S.L. Hsu, W.H. Haemmerle, B. Vis, C.K. Chau, and H. Weinstock, Solid State Comm. 15, 1185 (1974); W.J.M. de Jonge, C.H.W. Swüste, K. Kopinga, and K. Takeda, Phys. Rev. B12, 5858 (1975).
21. B. Vis, C.K. Chau, H. Weinstock and R.E. Dietz, Solid State Comm. 15, 1765 (1974).
22. H.W. White, K.H. Lee, J. Trainor, D.C. McCollum, and S.L. Holt, AIP Conf. Proc. # 18, 1974.
23. L.R. Walker, R.E. Dietz, K. Andres and S. Darack, Solid State Comm. 11, 593 (1972).
24. M.E. Fisher, Amer. J. Phys. 32, 343 (1964).
25. T. Smith and S.A. Friedberg, Phys. Rev. 176, 660 (1968).
The compound $CsMnCl_3 \cdot 2H_2O$ is one of the most extensively studied linear chain systems (1). Recent references include K. Kopinga, T. de Neef, and W.J.M. de Jonge, Phys. Rev. B11, 2364 (1975); W.J.M. de Jonge, K. Kopinga, and C.H.W. Swüste, Phys. Rev. B14, 2137 (1976); T. Iwashita and N. Uryû, J. Chem. Phys. 65, 2794 (1976).
26. C.Y. Weng, Tesis, Carnegie-Mellon University, 1969.
27. M.E. Harringan and G.L. Jones, Phys. Rev. B7, 4897 (1973).
28. H.W.J. Blöte, Physica 79B, 427 (1975).
29. T. de Neef and W.J.M. de Jonge, Phys. Rev. B11, 4402 (1975); T. de Neef, Phys. Rev. B13, 4141 (1976).
30. N. Achiwa, J. Phys. Soc. Japan 27, 561 (1969).
31. C. Dupas and J.-P. Renard, J. Chem. Phys. 61, 3871 (1974); K. Kopinga, T. de Neef, W.J.M. de Jonge, and B.C. Gerstein, Phys. Rev. B13, 3953 (1975).
32. J.F. Ackerman, G.M. Cole, and S.L. Holt, Inorg. Chim. Acta 8, 323 (1974).
33. S. Katsura, Phys. Rev. 127, 1508 (1962).

34. H.T. Witteveen and J. Reedijk, J. Solid State Chem. 10, 151 (1974); F.W. Klaaijsen, H. den Adel, Z. Dokoupil, and W.J. Huiskamp, Physica 79B, 113 (1975).
35. H.A. Algra, L.J. de Jongh, H.W.J. Blöte, W.J. Huiskamp, and R.L. Carlin, Physica 82B, 239 (1976); J.N. McElearney, S. Merchant, G.E. Shankle, and R.L. Carlin, J. Chem. Phys. 66, 450 (1977).
36. J.W. Metselaar, L.J. de Jongh, and D. de Klerk, Physica 79B, 53 (1975).
37. M. Date, M. Motokowa, H. Hori, S. Kuroda and K. Matsui, J. Phys. Soc. Japan 39, 257 (1975); W. Duffy, Jr., F.M. Weinhaws and D.L. Strandburg, Phys. Lett. 59A, 491 (1977).
38. J.N. McElearney, S. Merchant and R.L. Carlin, Inorg. Chem. 12, 906 (1973).
39. R.N. Rogers and C.W. Dempsey, Phys. Rev. 162, 333 (1967); S.N. Bhatia, C.J. O'Connor, R.L. Carlin, H.A. Algra and L.J. de Jongh, Chem. Phys. Lett.
40. L.S.J.M. Henkens, K.M. Diederix, T.O. Klaassen, and N.J. Poulis, Physica 81B, 259 (1976); L.S.J.M. Henkens, M.W. van Tol, K.M. Diederix, T.O. Klaassen, and N.J. Poulis, Phys. Rev. Lett. 36, 1252 (1976); L.S.J.M. Henkens, K.M. Diederix, T.O. Klaassen, and N.J. Poulis, Physica 83B, 147 (1974); L.S.J.M. Henkens, Thesis, Leiden, 1977.
41. J.N. McElearney, Inorg. Chem. 15, 823 (1976); R.E. Caputo, S. Roberts, R.D. Willett, and B.C. Gerstein, Inorg. Chem. 15, 820 (1976).
42. M. Steiner, J. Villain and C.G. Windsor, Adv. Phys. 25, 87 (1976).
43. L. Onsager, Phys. Rev. 65, 117 (1944).
44. R.F. Wielinga, H.W.J. Blöte, J.A. Roest, and W.J. Huiskamp, Physica 34, 223 (1967).
45. J.H.P. Colpa, Physica 57, 347 (1972); L.J. de Jongh, W.D. Van Amstel, and A.R. Miedema, Physica 58, 277 (1972); P. Bloembergen and A.R. Miedema, Physica 75, 205 (1974); P. Bloembergen, Physica 79B, 467 (1975); 81B, 205 (1976); L.J. de Jongh, Physica 82B, 247 (1976).
46. D.J. Breed, Physica 37, 35 (1967).
47. R. Navarro, J.J. Smit, L.J. de Jongh, W.J. Crama, and D.J.W. Ijdo, Physica 83B, 97 (1976); A.F.M. Arts, C.M.J. van Vyen, J.A. van Luijk, H.W. de Wijn and C.J. Beers, Solid State Comm. 21, 13 (1977).
48. E. Gurewitz, J. Makovsky, and H. Shaked, Phys. Rev. B14, 2071 (1976).
49. P.R. Newman, J.L. Imes, and J.A. Cowen, Phys. Rev. B13, 4093 (1976).
50. J.N. McElearney, H. Forstat, P.T. Bailey, and J.R. Ricks, Phys. Rev. B13, 1277 (1976).
51. J.P.A.M. Hijmans, W.J.M. de Jonge, P. van der Leeden, and M.J. Steenland, Physica 69, 76 (1973); K. Kopinga, P.W.M. Borm and W.J.M. de Jonge, Phys. Rev. B10, 4690 (1974); J.P.A.M. Hijmans, Q.A.G. van Vlimmeren, and W.J.M. de Jonge, Phys. Rev. B12, 3859 (1975).
52. J.J. White and S.N. Bhatia, J. Phys. C 8, 1227 (1975).
53. J.A.J. Basten and A.L.M. Bongaarts, Phys. Rev. B14, 2119 (1976).
54. J. Mizuno, J. Phys. Soc. Japan 15, 1412 (1960).
55. M.E. Lines, Phys. Rev. B12, 3766 (1975).
56. M.E. Fisher, Proc. Roy. Soc. (London) A254, 66 (1960); M.E. Fisher and M.F. Sykes, Physica 28, 939 (1962).

CHAPTER VII

SPECIAL TOPICS:
SPIN-FLOP, METAMAGNETISM, FERRIMAGNETISM AND CANTING

A. INTRODUCTION

Most of what has been presented above concerns either zero-field behavior of ordered systems, or else paramagnetic systems where the magnetization depends linearly on the magnetic field. A number of interesting new effects occur upon the application of a magnetic field to an ordered system, and several of these effects will be described in this chapter. Furthermore, spin structures are frequently more complicated than the simple lattice models presented in Chapter V, and several interesting examples will be presented here. The discussion of this chapter is restricted to temperatures below T_c.

One simple example will illustrate the sensitivity of magnetically-ordered systems to a magnetic field, and that is that an applied external field will destroy a ferromagnetic phase transition. This occurs because, in a ferromagnet, the spins are already spontaneously aligned in a preferential domain arrangement in zero-field that results in a large magnetization. Application of even a weak field causes the magnetization to follow the direction of the field, thus destroying the transition. A phase diagram can be constructed for ordered magnetic systems in the H-T plane, to illustrate the several phases that occur. Since a ferromagnet can only be found in zero external field, the H-T phase diagram simply consists of a line of points on the T axis, terminating at $T = T_c$.

B. PHASE DIAGRAMS AND SPIN-FLOP (1-3)

The phenomenon of spin-flop is observed in the study of the behavior of certain antiferromagnets in the H-T plane. Thus, the discussion so far of the ordering temperature consists of assigning merely one point on a graph in which T is plotted as the abscissa and external H as the ordinate; that point, $T_c(H=0)$, lies on the T axis at H = 0. Upon application of a field parallel to the preferred axis of magnetization of a antiferromagnet, a competition is set up between the strength of the external field and that of the exchange interactions, expressed as an internal exchange field. In the case of an antiferromagnet with weak anisotropy, when H reaches a critical field H_{SF}, the anti-parallel magnetizations of the two sublattices turn (flop) from the direction of the easy axis to that perpendicular to it. This so-called spin flop phase transition was first observed in the early '50's by the Leiden group (4), though it was predicted earlier by Néel.

Once again, the behavior of a system in a magnetic field depends on the dimensionality and anisotropy of the system. The planar square Ising antiferromagnet has a particularly simple phase diagram, as is illustrated in Figure 7.1 for Cs_3CoBr_5 (5), because the spin-flop phase does not appear in a system with the large anisotropy characterized by the Ising model. This diagram was determined by means of two kinds of experiments. In the first, the specific heat was measured in various external magnetic fields and, as illustrated in Figure 7.2, the maxima in these curves shift to lower temperatures as the field increases. These data are plotted as the boxes in Figure 7.1. The remaining points were taken from the isentropes, which are determined in the following fashion. The sample is thermally isolated in zero field at some temperature $T < T_c$, and then the magnetic field is increased quickly (say, from 0 to 2500 Oersteds in 150 sec.) and the temperature monitored. A smooth curve is obtained of T vs. H, with a minimum, and it was assumed (5) that the minimum corresponds to a point on the phase boundary (circles in Fig. 7.1). As the boundary is crossed from the AF side, the spins are completely aligned and, though this corresponds to a saturated paramagnetic state, it is sometimes called a ferromagnetic state. Thus, at T = 0, the spins become ferromagnetically aligned at a critical external field $H_c = H_E$, where H_E is called an antiferromagnetic exchange field. That is, the exchange energy of $2z|J|S$ may be equated to an energy expressed in terms of a field, $g\mu_B H_E$. By raising the temperature, the critical field $H_c(T)$ decreases continuously until it vanishes at the critical point, $T_c(H=0)$. Alternatively, one can say that the critical temperature $T_c(H)$ decreases with increasing field.

Turn now to the field dependent behavior of a Heisenberg antiferromagnet with small exchange anisotropy. In contrast to the Ising example

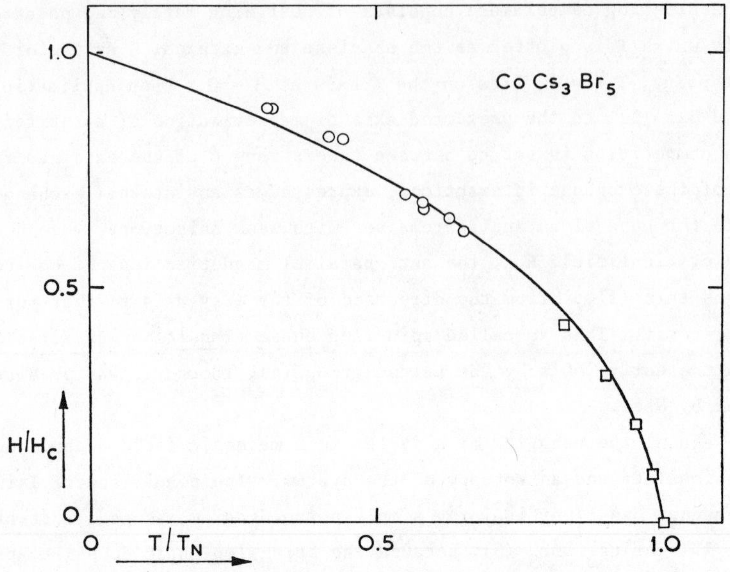

FIGURE 7.1 *Phase diagram of Cs_3CoBr_5 on a reduced temperature scale, taking H_c = 2100 Oe at T = 0 K. The drawn line is computed for a planar square Ising antiferromagnet. From Ref. 5.*

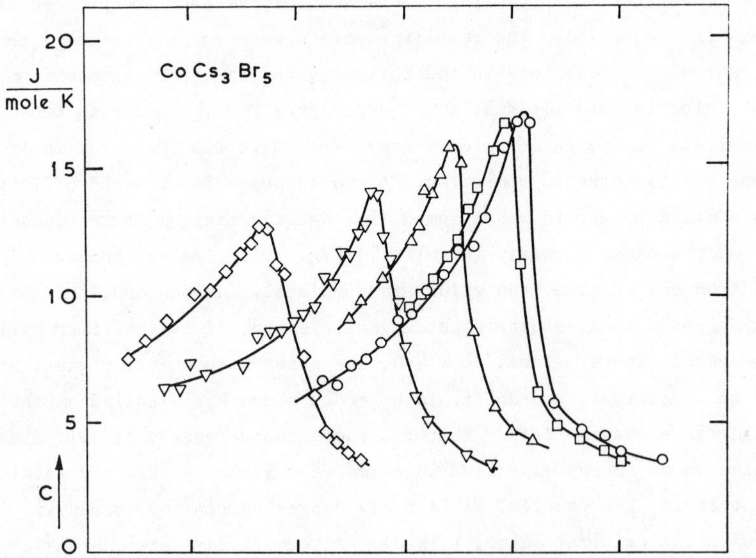

FIGURE 7.2 *Heat capacity curves of Cs_3CoBr_5 in various magnetic fields: ○ 0 Oe, □ 220 Oe, △ 440 Oe, ▽ 660 Oe, ◇ 880 Oe. From Ref. 5.*

cited above, this system displays the phenomenon of spin-flop over a certain range of fields and temperatures below T_c. The spin-flop transition is most easily described by considering a simple uniaxial two-sublattice molecular field model at T = 0 K. There is a certain anisotropy energy K, which may also be expressed in terms of a field H_A, which causes the establishment of the preferred or easy axis in some particular crystallographic direction. For example, one of the contributors to H_A is the single-ion anisotropy associated with a zero-field splitting. The free energy of an antiferromagnet in a field H is proportional to $-\frac{1}{2}\chi_{\parallel}H^2$ for fields parallel to the easy axis, and to $K - \frac{1}{2}\chi_{\perp}H^2$ in case the external field is perpendicular to the preferred axis. In the usual antiferromagnet, of course, $\chi_{\perp}$ is always larger than $\chi_{\parallel}$ so that if an external field is applied parallel to the preferred axis of antiferromagnetic alignment, the moments will have a tendency to orient themselves perpendicular to the field. In so doing, they gain the magnetic energy $\frac{1}{2}(\chi_{\perp} - \chi_{\parallel})H^2$. In small external fields the anisotropy field H_A will exceed the latter term, but at a certain critical field when $-\frac{1}{2}\chi_{\parallel}H^2 = K - \frac{1}{2}\chi_{\perp}H^2$ or

$$H_{SF} = [2K/(\chi_{\perp} - \chi_{\parallel})]^{\frac{1}{2}} \tag{7.1}$$

the spins will flip over to the perpendicular orientation. A further increase of field will gradually rotate the sublattice moments, until at the critical field H_c their mean direction is parallel to the field (and the easy axis) and the paramagnetic phase is entered. At this point on the SF-P boundary, the antiferromagnetic interaction is balanced by the applied field and anisotropy field (1). The spin-flop phase is naturally absent if the field is applied perpendicular to the preferred axis.

The behavior of the magnetization and of the differential susceptibility as a function of field expected on the basis of this model are sketched in Figure 7.3 In this molecular field model, note that there is a discontinuity in the magnetization at the spin-flop transition, but that the magnetization is continuous while the susceptibility is discontinuous at the SF-P transition. It is common to classify the phase transition as first-order (discontinuity in M) or as second order (M is continuous, but $\chi = dM/dH$ is not.) Between the two transition fields H_{SF} and H_c in the figure, the magnetization increases linearly with H according to

$$M/M_S = H/(2H_E - H_A) \tag{7.2}$$

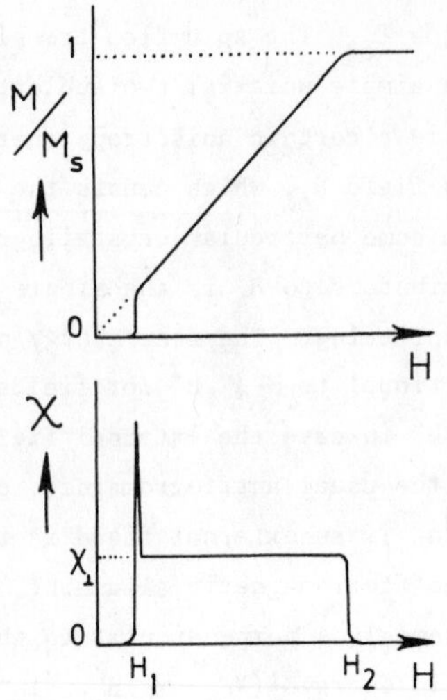

FIGURE 7.3

The behavior as a function of field of the isothermal magnetization and differential susceptibility of a weakly anisotropic antiferromagnet, according to the MF theory for a temperature near T = 0 K. The fields H_1 and H_2 correspond to the spin-flop field H_{SF} and the transition from the flopped to the paramagnetic phase (H_c), respectively. From Ref. 3.

where M_S is the saturation magnetization of each sublattice $\frac{1}{2}Ng\mu_B S$, H_E was introduced above as the antiferromagnetic exchange field, and H_A is the anisotropy field. The formulas for the critical fields at T = 0 K are (3)

$$H_{SF}(0) = (2H_E H_A - H_A^2)^{1/2} \tag{7.3}$$

$$H_c(0) = 2H_E - H_A \tag{7.4}$$

These critical fields are dependent on temperature and a theoretical phase diagram is shown in Figure 7.4. The fields $H_{SF}(T)$ and $H_c(T)$ are usually obtained over as wide a temperature interval as possible, and the zero-temperature values obtained by extrapolation. Furthermore, the perpendicular susceptibility at T = 0 K can be written as

$$\chi_\perp(0) = 2M_S/(2H_E + H_A), \tag{7.5}$$

so that a smooth extrapolation of $\chi_\perp(T)$ to T = 0 K, along with Eqs. (7.3-5), provides a way of estimating H_E and H_A. Since there are two sublattices, one uses $M_S = Ng\mu_B S/2$. Determination of H_E allows an estimate of the exchange constant to be made, for molecular field theory also results in the relationship

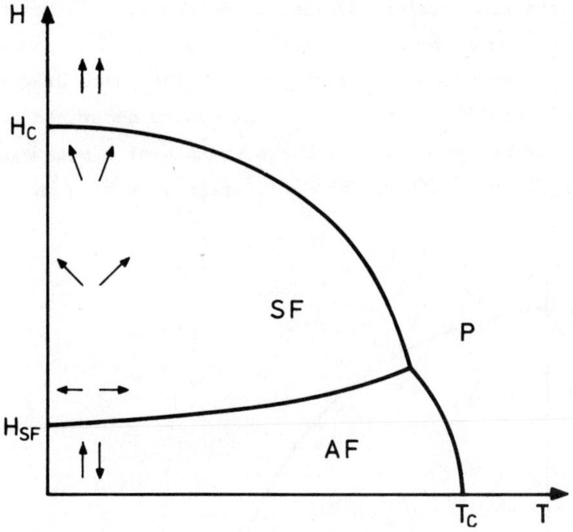

FIGURE 7.4 *Theoretical magnetic phase diagram of a weakly anisotropic antiferromagnet.*

$$H_E = 2z|J|S/g\mu_B. \tag{7.6}$$

Notice that the AF-SF and SF-P phase boundaries meet at a triple point, called the bicritical point, together with the AF-P transition curve. The behavior of the susceptibilities near this point is one of the areas of present theoretical and experimental studies.

Manganese chloride tetrahydrate, $MnCl_2 \cdot 4H_2O$, is one of the most thoroughly investigated antiferromagnets. Its phase diagram, illustrated in Figure 7.5, has been obtained by a variety of procedures, such as isentropic magnetization, susceptibility, and specific heat measurement in various external fields (6). Some of the latter data are illustrated in Figure 7.6. It should be obvious that these experiments are necessarily carried out on oriented single crystals.

C. METAMAGNETISM (1,3)

A metamagnetic system is one in which, because of important ferromagnetic interactions, the spin-flop region of the relevant phase diagram has shrunk to zero area. From Eqs. (7.3) and (7.4) above, one can observe that an increase in anisotropy increases H_{SF} while it lowers H_c. When $H_A = H_E$, then the two critical fields become equal, and the anisotropy is so large that the moments go over directly from an AF alignment (parallel to H) to a saturated paramagnetic alignment. The phase diagram then resembles, coincidentally, that of the Ising system described above. The compound $FeCl_2$ is perhaps the best-known example of a metamagnet. The crystal and magnetic structures of $FeCl_2$ are illustrated in Figure 7.7,

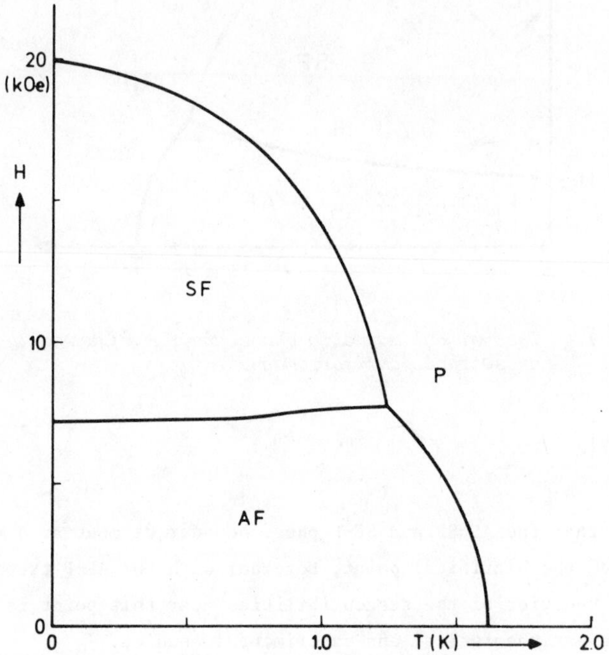

FIGURE 7.5 *Experimental antiferromagnetic phase diagram as obtained by Giauque for* $MnCl_2 \cdot 4H_2O$ *(6). H is the applied field.*

where the compound will be seen to consist of layers of iron atoms, separated by chloride ions. The coupling between the ions within each layer is ferromagnetic, which is the source of the large anisotropy energy that causes the system to behave as a metamagnet, and the layers are weakly coupled antiferromagnetically. Because of the large anisotropy, the transition from the AF to paramagnetic phase occurs in the fashion described above with, for $T \ll T_c$, a discontinuous rise of the magnetization at the transition field to a value near to saturation. The isothermal magnetiz-

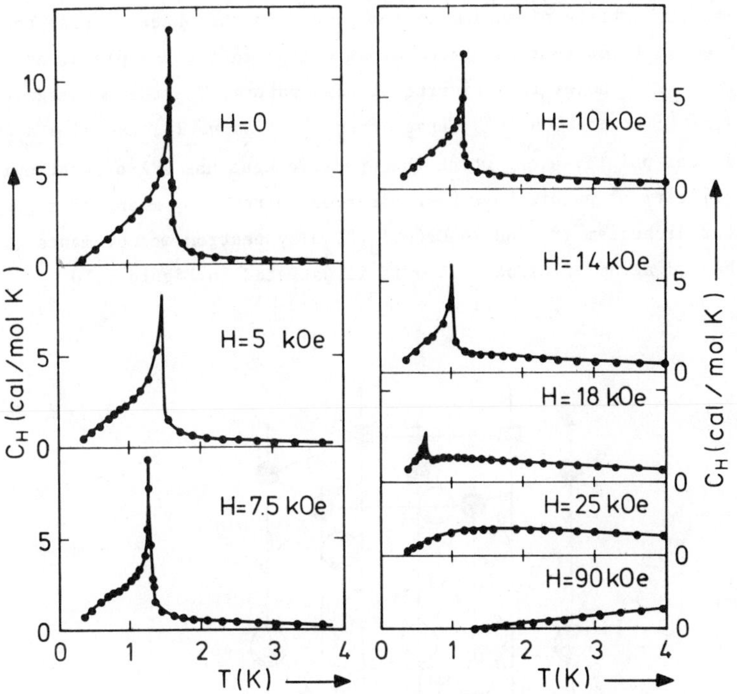

FIGURE 7.6 *The specific heat of $MnCl_2 \cdot 4H_2O$ in various (constant) applied fields. After Reichert and Giauque, Ref. 6.*

ations illustrated in Figure 7.8 are qualitatively different in behavior from those in Figure 7.3. The phase diagram that results is also given in Figure 7.8, where it will also be noted that the boundary above about 20.4 K is dashed rather than solid. This is because the tricritical point of $FeCl_2$ occurs at about 20.4 K.

A tricritical point is found to occur in the phase diagram with metamagnets as well as in certain other systems, such as ^3He - ^4He (3,7). The metamagnet must have important ferromagnetic interactions, even though the overall magnetic structure will be antiferromagnetic. In the H-T plane, then, although a continuous curve is observed as in Figure 7.8, it consists of a set of first-order points at low temperature which, at the tricritical point, goes over into a set of second-order points. It arises upon consideration of the addition of a third dimension to the H-T phase diagram of a metamagnet, the third dimension being a staggered field, $\tilde{H}$.

This is an experimentally unattainable field that alternates in direction with each sublattice of an antiferromagnet. In the phase diagram there are three surfaces that intersect along a line in the H-T plane, as shown in Figure 7.9. Below the tricritical temperature, T_t, the metamagnetic-paramagnetic transition is a first-order one. Above T_t, there is a λ-line of critical points, along which the specific heat has a λ-discontinuity. Such tricritical points have been observed in $FeCl_2$ by means of magnetic circular dichroism (8) and in $CsCoCl_3.2D_2O$ by neutron measurements (9). The phase diagram of $CsCoCl_3.2D_2O$ is illustrated in Figure 7.10.

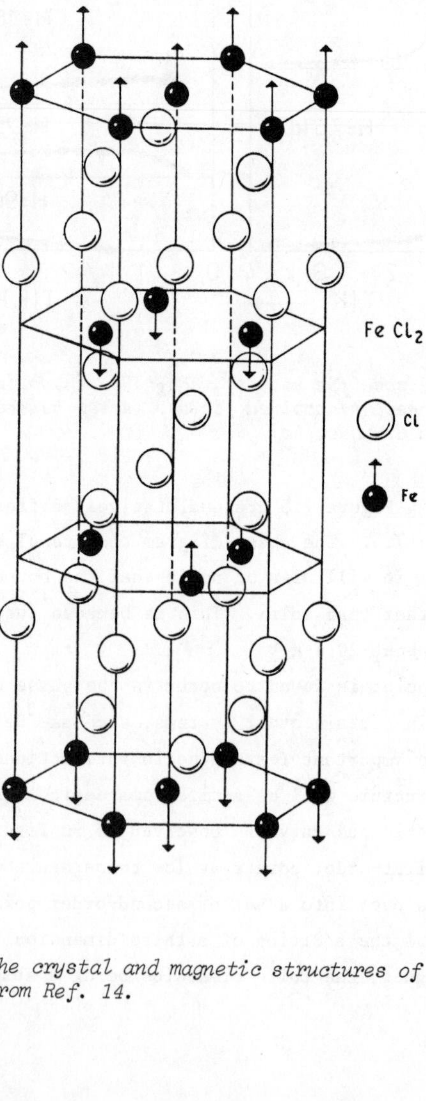

FIGURE 7.7 *The crystal and magnetic structures of* $FeCl_2$. *From Ref. 14.*

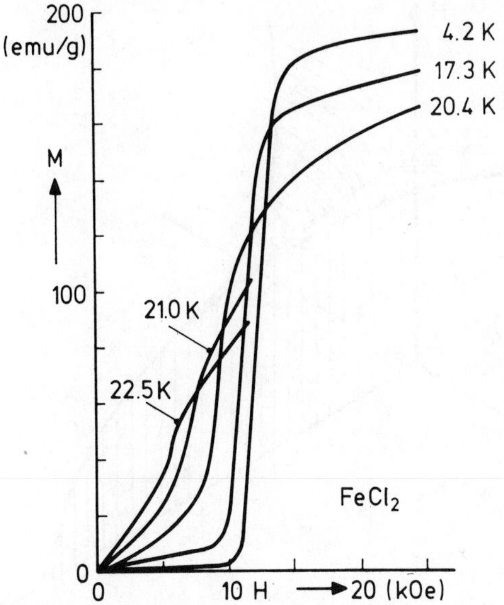

(a)

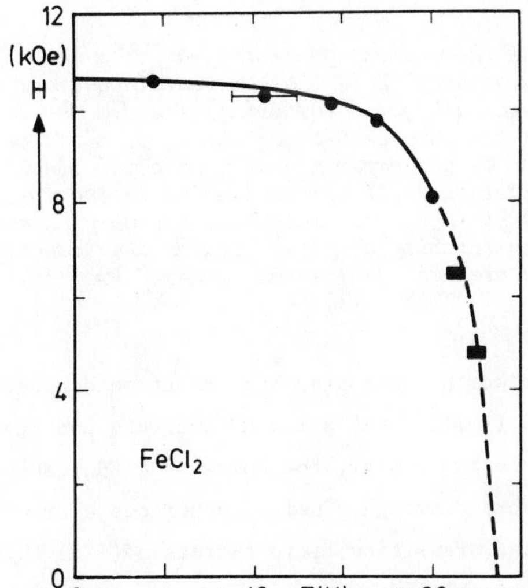

(b)

FIGURE 7.8 *(a) Magnetization isotherms of the metamagnet* $FeCl_2$ *as measured by Jacobs and Lawrence (1967). The transition temperature is 23.5 K.*
(b) The metamagnet phase diagram of $FeCl_2$ *showing the first-order transition line (solid curve) and the second-order line (dashed curve). The tricritical point is estimated to be at 20.4 K. From Ref. 3.*

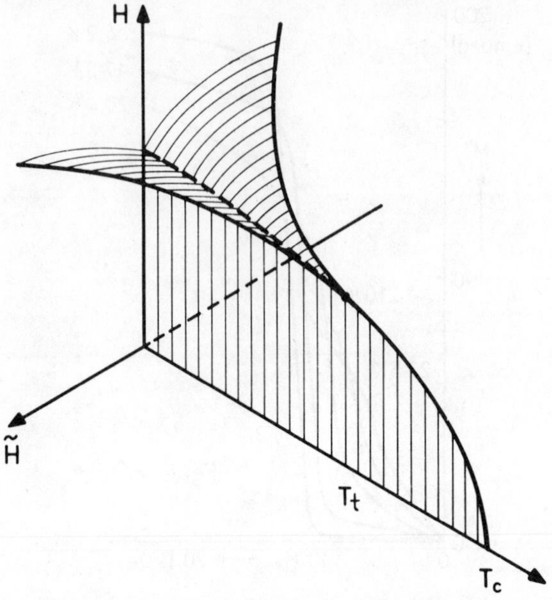

FIGURE 7.9 *The theoretical phase diagram of a metamagnetic substance. H is the constant (non-ordering) field, $\tilde{H}$ the staggered (ordering) field. The dashed line is the intersection of three co-existence surfaces and is the experimentally observed line of first order transitions terminating in the tricritical point (H_t, T_t). The phase boundary extending from the tricritical point towards the temperature axis is presumably of second order. From Ref. 3.*

One last point worth mentioning here about metamagnets is that it has recently been found (10,11) that a few metamagnets undergo transitions at quite low fields. In particular, the compound $[(CH_3)_3NH]CoCl_3 \cdot 2H_2O$, which will be discussed more thoroughly below, undergoes a transition at 64 Oersteds (10), while the transition field increases (11a) to 120 Oersteds when bromide replaces chloride in this compound. More impressively, $Mn(OAc)_2 \cdot 4H_2O$ undergoes a transition at a mere 6 Oersteds (11b)! The reason why the transition fields are so low in these compounds is not clear yet, but they all have a crystal lattice in which substantial hydrogen bonding occurs, and they all exhibit canting (Section E, below). Certainly these results suggest that caution be applied by doing experiments in zero field before they are carried out in the presence of a large external field.

D. FERRIMAGNETISM (1,12)

We have suggested up until now that ordered substances have spins which are ordered exactly parallel (ferromagnets) or antiparallel (antiferromagnets) at 0 K. This is not an accurate representation of the true situation, for we are generally ignoring in this book the low energy excitations of the spin systems which are called spin-waves or magnons (3). Other situations also occur, such as when the antiparallel lattices are not equivalent, and we briefly mention one of them in this section, the subject of ferrimagnetism. Most ferrimagnets studied to date are oxides of the transition metals, such as the spinels (13) and the garnets (14). But there is no apparent reason why transition metal complexes of the kind discussed in this book could not be found in the future to display ferrimagnetism.

Ferrimagnetism is found when the crystal structure of a compound is more complicated than has been implied above, so that the magnitudes of the

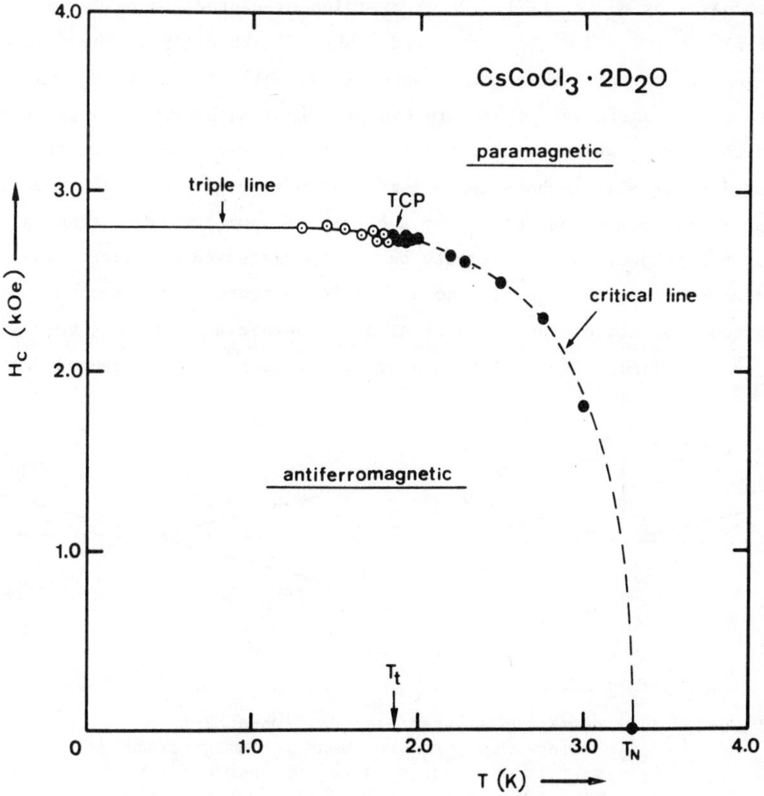

FIGURE 7.10 *Magnetic phase diagram of $CsCoCl_3 \cdot 2D_2O$. From Ref. 9.*

magnetic moments associated with the two AF sublattices are not exactly the same. Then, when the spontaneous anti-parallel alignment occurs at some transition temperature, the material retains a small but permanent magnetic moment, rather than a zero one. The simplest example is magnetite, Fe_3O_4, a spinel. The two chemical or structural sublattices are 1) iron(III) ions in tetrahedral coordination to oxygen, and 2) iron(II) and iron(III) ions in equal proportion to octahedral oxygen coordination. The result is that the inequivalent magnetic sublattices cannot balance each other out, and a weak moment persists below T_c.

E. CANTING AND WEAK FERROMAGNETISM (1,15)

Certain other substances that are primarily antiferromagnetic exhibit a weak ferromagnetism that is due to an entirely different physical phenomenon, a canting of the spins. This was first realized by Dzyaloshinsky (16) in a phenomenological study of α-Fe_2O_3, and later put on a firm theoretical basis by Moriya (17). Other examples of canted compounds include NiF_2 (17), $CsCoCl_3 \cdot 2H_2O$ (18), and $[(CH_3)_3NH]CoCl_3 \cdot 2H_2O$ (19). The compound $CoBr_2 \cdot 6D_2O$ is apparently a canted magnet (20), while the hydrated analogue is not. As we shall see, there are two principal mechanisms, quite different in character, which cause canting, but there is a symmetry restriction that applies equally to both mechanisms. In particular, magnetic moments in a unit cell cannot be related by a center of symmetry if canting is to occur. Other symmetry requirements have been discussed elsewhere (15).

Weak ferromagnetism is due to an antiferromagnetic alignment of spins on the two sublattices that is equivalent in number and kind but not exactly antiparallel. The sublattices may exist spontaneously canted, with

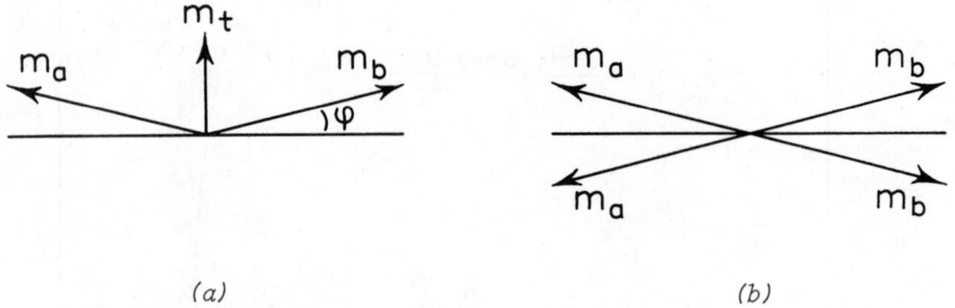

FIGURE 7.11 *Canting of sublattices: (a) two-sublattice canting which produces weak ferromagnetism; (b) "hidden" canting of a four-sublattice antiferromagnet. Four sublattices may also exhibit "overt" canting, with a net total magnetic moment. From Ref. 1.*

no external magnetic field, only if the total symmetry is the same in the canted as in the uncanted state. The magnetic and chemical unit cells must be identical. The canting angle, usually a matter of only a few degrees, is of the order of the ratio of the anisotropic to the isotropic interactions. If only two sublattices are involved, the canting gives rise to a small net moment, m_t, and weak ferromagnetism occurs. This is sketched in Figure 7.11a. If many sublattices are involved, the material may or may not be ferromagnetic, in which case the canting is called, respectively, overt or hidden. Hidden canting is illustrated in Figure 7.11b, and just this configuration has indeed been found to occur at zero field in $LiCuCl_3 \cdot 2H_2O$ (21). It has also been suggested that hidden canting occurs in $CuCl_2 \cdot 2H_2O$ (17).

Another example is provided by $CsCoCl_3 \cdot 2H_2O$, whose molecular structure is illustrated in Figure 7.12. Chains of chloride-bridged cobalt atoms run along parallel to the a axis, and the compound exhibits a high degree of short-range order (18). The proposed magnetic structure is illustrated in Figure 7.13. This spin configuration was obtained from an analysis of the nuclear resonance of the hydrogen and cesium atoms in the ordered state (i.e., $T < T_c$). Notice that the moments are more-or-less AF aligned along the chains (a axis), but that they make an angle ϕ with the

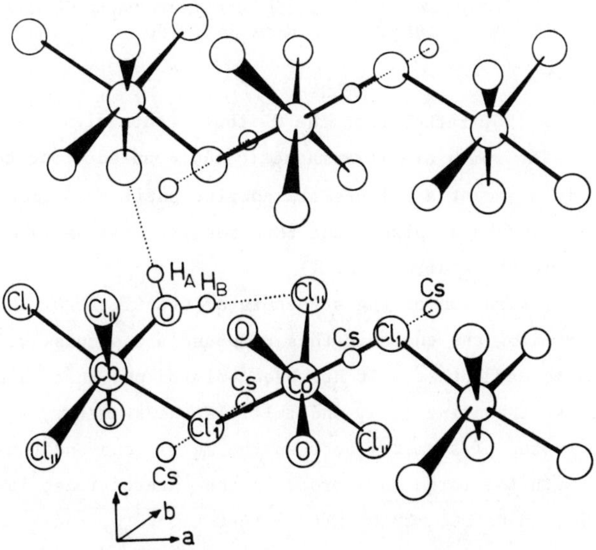

FIGURE 7.12 *Structure of $CsCoCl_3 . 2H_2O$, according to Thorup and Soling. Only one set of hydrogen atoms and hydrogen bonds are shown. From Ref. 18.*

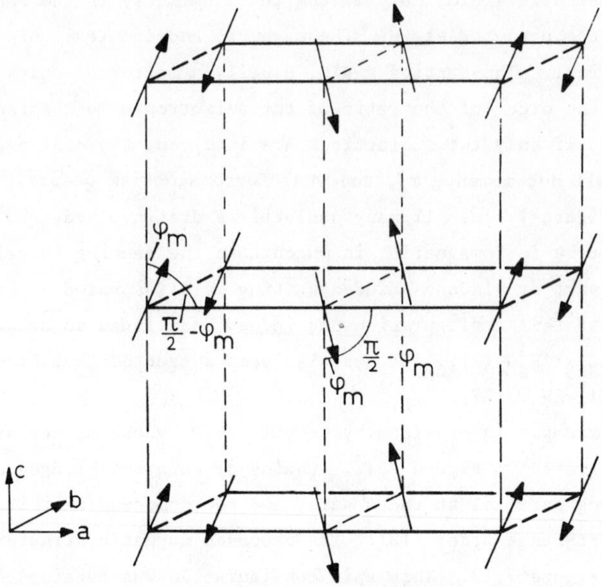

FIGURE 7.13 *Proposed magnetic-moment array of $CsCoCl_3 \cdot 2H_2O$ (18). All spins lie in the ac plane. The model suggests ferromagnetic coupling along the c axis, antiferromagnetic coupling along the b axis, and essentially antiferromagnetic coupling along the a axis. From Ref. 18.*

c axis of some 15°, rather than the 0° that occurs with a normal antiferromagnet. The spins are ferromagnetically coupled along the c axis, but moments in adjacent ac planes are coupled antiferromagnetically. The spins thus lie in the ac plane, and this results in a permanent, though small, moment in the a direction.

As the illustration of the structure of $CsCoCl_3 \cdot 2H_2O$ clearly shows, the octahedra along the chain in this compound are successively tilted with respect to each other. It has been pointed out that this is an important source of canting (22), and fulfills the symmetry requirement of a lack of a center of symmetry between the metal ions and thus the moments. This, along with the large anisotropy in the interactions, is the source of the canting in $CsCoCl_3 \cdot 2H_2O$, $[(CH_3)_3NH]CoCl_3 \cdot 2H_2O$, and α-$CoSO_4$ (22).

Now the Hamiltonian we have used repeatedly to describe exchange,

$$\mathcal{H} = -2J\, \vec{S}_i \cdot \vec{S}_j \tag{7.7}$$

may be said to describe symmetric exchange, causes a normal AF ordering of spins, and does not give rise to a canting of the spins. The Hamiltonian

$$\mathcal{H} = \vec{D}_{ij} \cdot [\vec{S}_i \times \vec{S}_j] \tag{7.8}$$

where $\vec{D}_{ij}$ is a constant vector, may then be said to describe what is called antisymmetric exchange. This latter Hamiltonian is usually referred to as the Dzyaloshinsky-Moriya (D-M) interaction. This coupling acts to cant the spins because the coupling energy is minimized when the two spins are perpendicular to each other. Moriya (15) has provided the symmetry results for the direction of $\vec{D}$, depending on the symmetry relating two particular atoms in a crystal. The more anisotropic the system, the more important canting will be, for $\vec{D}$ is proportional to $(g-2)/g$. In $CsCoCl_3 \cdot 2H_2O$, with $g_a = 3.8$, $g_b = 5.8$ and $g_c = 6.5$, the required anisotropy is obviously present, and $(g-2)/g$ can be as much as 0.7.

The geometry of β-MnS allows the D-M interaction to arise, and Keffer (23) has presented an illuminating discussion of the symmetry aspects of the problem. Cubic β-MnS has the zinc blende structure. The lattice is composed of an fcc array of manganese atoms interpenetrating an fcc array of sulfur atoms in such a way that every atom of one kind is surrounded tetrahedrally by atoms of the other kind. The local symmetry at one sulfur atom is displayed in Figure 7.14; if the midpoint of a line connecting the manganese atoms were an inversion center, that is, if another sulfur atom were present below the line of cation centers, then the D-M mechanism

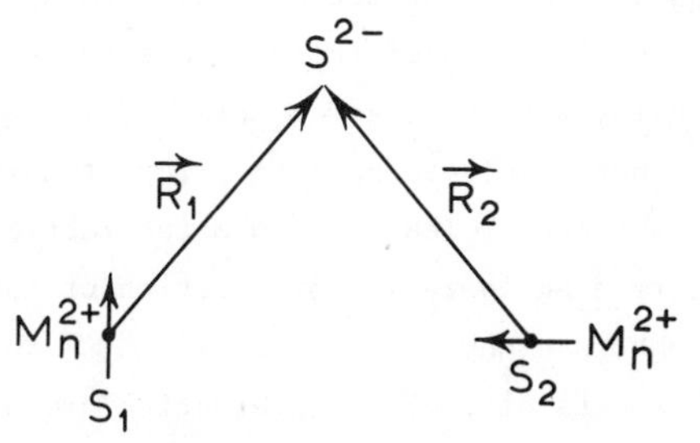

FIGURE 7.14 *Near-neighbor superexchange in β-MnS. The absence of an anion below the line of cation centers allows a Moriya interaction, with D taking the direction of* $\vec{R}_1 \times \vec{R}_2$. *From Ref. 23.*

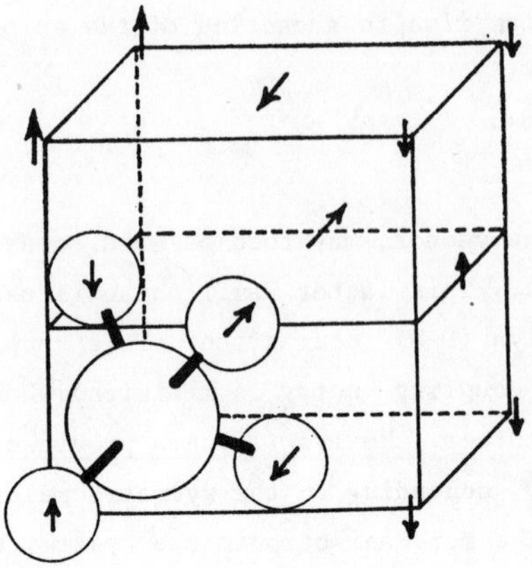

FIGURE 7.15 *Cubic cell of β-MnS. Only the sulfur in the lowest left front corner is shown. The proposed arrangement of manganese spins is indicated by the arrows. From Ref. 23.*

would not be allowed. As it is, the absence of such an anion, even though the total crystal symmetry is high, allows such a D-M coupling. The direction of $\vec{D}_{12}$ is normal to the plane of the paper, in the direction of the vector $\vec{R}_1 \times \vec{R}_2$, and Keffer finds that the energy associated with the D-M interaction will be minimum when $\vec{S}_1$ and $\vec{S}_2$ are orthogonal, as in the figure. Furthermore, Keffer goes on to analyze the spin structure of the lattice and concludes that the probable spin arrangement, illustrated in the Figures 7.15 and 7.16, is not only consistent with the available neutron diffraction results, but is in fact determined primarily by the D-M interaction. Spins lie in an antiferromagnetic array in planes normal to the x axis. The spin direction of these arrays turns by 90° from plane to plane, the x axis being a sort of screw axis. Next-neighbor superexchange along the x axis is not likely to be large, for it must be routed through two intervening sulfur ions. Thus, the D-M interaction alone is capable of causing the observed antiparallel orientation of those manganese ions which are next neighbors along x.

The symmetry aspects of the D-M interaction are also illustrated nicely by the spin structure of α-Fe_2O_3 and Cr_2O_3, as was first suggested by Dzyaloshinsky (16). The crystal and magnetic structures of these isomorphous crystals will be found in Figure 7.17, and the spin arrangements of the two substances will be seen to differ slightly. Spins of the ions 1,2,3, and 4 differ in orientation only in sign and their sum in each unit

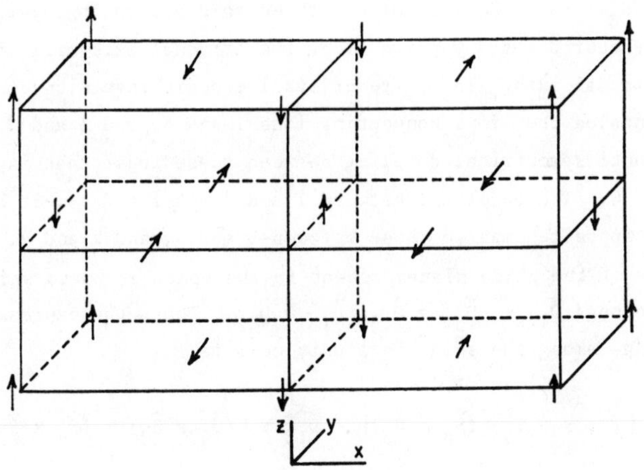

FIGURE 7.16 *Proposed spin arrangement in cubic β-MnS. From Ref. 23.*

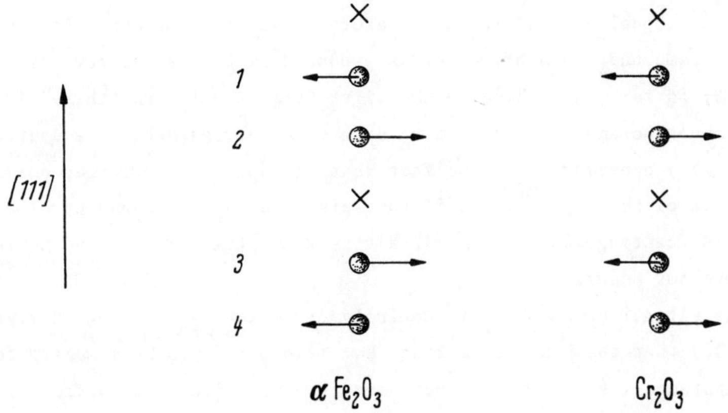

FIGURE 7.17 *Arrangement of spins along c axis in unit cells of α-Fe_2O_3 and of Cr_2O_3 (schematic). Points marked × are inversion centers of the crystallochemical lattices.*

cell is equal to zero. In α-Fe_2O_3, $\vec{S}_1 = -\vec{S}_2 = -\vec{S}_3 = \vec{S}_4$, while in Cr_2O_3, $\vec{S}_1 = -\vec{S}_2 = \vec{S}_3 = -\vec{S}_4$. Because of the three-fold crystallographic rotation axis, the vector $\vec{D}$ for any pairs along the trigonal axis will lie parallel to that trigonal axis. There are crystallographic inversion centers at the midpoints on the lines connecting ions 1 and 4, and 2 and 3, so there cannot be anti-symmetrical coupling between these ions; that is, $\vec{D}_{23} = \vec{D}_{14} = 0$. The couplings between 1 and 2 and 1 and 3 must be equal but of the opposite sign to those between 3 and 4, and 2 and 4, respectively, because of the glide plane present in the space group to which those crystals belong: $\vec{D}_{12} = -\vec{D}_{34} = \vec{D}_{43}$; $\vec{D}_{13} = \vec{D}_{42}$. Thus, the expression for the couplings among the spins in a unit cell is

$$\vec{D}_{12} \cdot [(\vec{S}_1 \times \vec{S}_2) + (\vec{S}_4 \times \vec{S}_3)] + \vec{D}_{13} \cdot [(\vec{S}_1 \times \vec{S}_3) + (\vec{S}_4 \times \vec{S}_2)] \qquad (7.9)$$

But, the spin arrangements in these crystals are

$$\alpha\text{-}Fe_2O_3 : \vec{S}_1 \parallel \vec{S}_4 \text{ and } \vec{S}_2 \parallel \vec{S}_3$$

$$Cr_2O_3 : \vec{S}_1 \parallel \vec{S}_3 \text{ and } \vec{S}_2 \parallel \vec{S}_4$$

For the iron compound, the cross products do not vanish, there is an anti-symmetrical coupling, and it is therefore a weak ferromagnet. For the chromium compound, each of the cross products in the second term of Eq. (7.9) is zero, and those in the first term cancel each other. Thus, Cr_2O_3 cannot be and is found not to be a weak ferromagnet. The canting in α-Fe_2O_3 is a consequence of the fact that the spins are arranged normal to the c axis of the crystal, for if the spins had been arranged parallel to this axis, canting away from [111] alters the total symmetry and could therefore not occur.

The situation with NiF_2 is entirely different, and in fact Moriya shows (15) that the D-M interaction, Eq. (7.8), is zero by symmetry for this substance. But, on the other hand, the nickel(II) ion suffers a tetragonal distortion which results in a zero-field splitting. For the case where there are, say, two magnetic ions in a unit cell and the AF ordering consists of two sublattices, one can write the single-ion anisotropy energies at the two positions as $E_1(\vec{S})$ and $E_2(\vec{S})$. If it is found that $E_1(\vec{S}) = E_2(\vec{S})$, as would happen if the spin-quantization or crystal field axes of the two ions were parallel, Moriya finds that a canted spin arrangement cannot be obtained because the preferred directions of the two

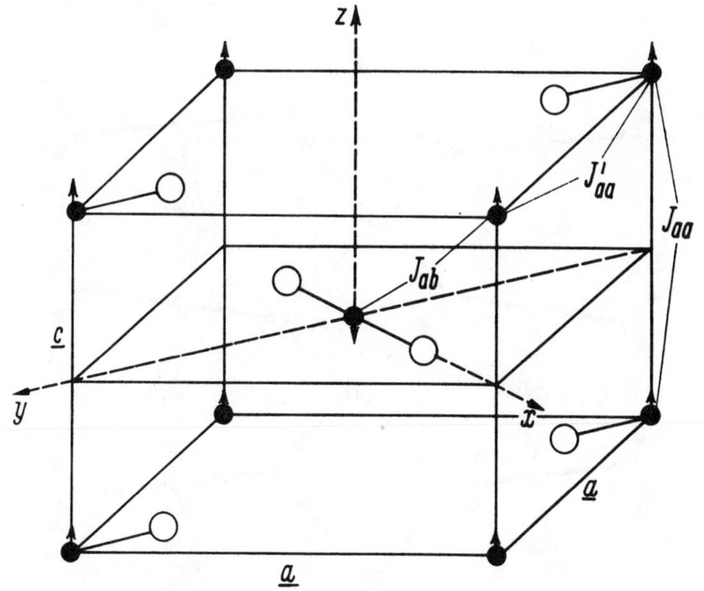

FIGURE 7.18 *Antiferromagnetism in MnF_2. Shown is the unit cell of this rutile-structure crystal. Open circles, F^- ions; closed circles, Mn^{++} ions, indicating spin directions below T_c.*

spins are then the same and the AF ordering should be formed in this direction. When $E_1(\vec{S}) \neq E_2(\vec{S})$, however, and the easiest directions for the spins at the two positions are different, canting of the sublattice magnetizations may take place. The crystal and magnetic structures (1) of tetragonal MnF_2 (as well as FeF_2 and CoF_2) are illustrated in Figure 7.18, where it will be seen that the c axis is the preferred spin direction and there is no canting. Any canting away from c would change the symmetry, in violation of the Dzyaloshinsky symmetry conditions. In NiF_2, however, the sublattices lie normal to the c axis and cant within the xy or ab plane. In this geometry there is no change of symmetry on canting.

Experimental results are in accordance (24). Because of the weak ferromagnetic moment, and in agreement with Moriya's theory, the paramagnetic susceptibility in the ab (⊥) plane rises very rapidly, becoming very large as T_c (73.3 K) is approached. This is of course contrary to the behavior of a normal antiferromagnet. The behavior follows not so much from the usual parameter D of the spin-Hamiltonian (which nevertheless must be

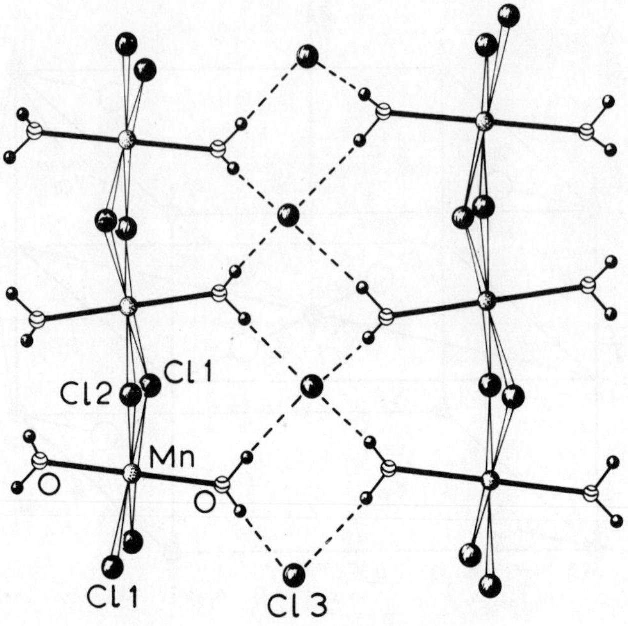

FIGURE 7.19 *The hydrogen-bonded plane of chains in* $[(CH_3)_3NH]MnCl_3 \cdot 2H_2O$. *From Ref. 26.*

positive) but from a competition between the exchange parameter J and the rhombic zero-field splitting term, E. This is the term that causes the spins on different lattice sites to be perpendicular. The total energy is minimized through a compromise in which the spins are tilted away from the antiparallel towards a position in which they are perpendicular. The canting angle is not large, however, being only of the order of a degree.

This mechanism may also be one of the sources of the canting observed in $[(CH_3)_3NH]MnBr_3 \cdot 2H_2O$ (25), where, as usual, the g-value anisotropy of manganese(II) is so small as to make the existence of an important D-M interaction unlikely. Large zero-field splittings are suggested by the analysis of the specific heat, however, and, as illustrated by the structure (26) of the chlorine analogue in Figure 7.19, the adjacent octahedra are probably tilted, one with respect to the next, and so the principal axes of the distortion meet the same symmetry requirements illustrated above by NiF_2.

In this regard, it is interesting to note that, although $CsCoCl_3 \cdot 2H_2O$ exhibits canting, the isomorphous $CsMnCl_3 \cdot 2H_2O$ does not (27). By analogy with the cobalt compound, the requisite symmetry elements that do not allow canting must be absent but apparently, both g-value and zero-field anisotropies are either too small to cause observable weak ferromagnetism, or else are zero. There is also no evidence to suggest that hidden canting occurs. In the discussions of β-MnS (23,28),,the sources of anisotropy considered were the D-M interaction and magnetic dipole anisotropy. Since that crystal is cubic, zero-field splittings are expected to be small, and g-value anisotropy is small. In those cases involving manganese where the zero-field splitting may be large, such as mentioned above, and the g-value anisotropy still small, this may be a more important factor in causing canting. It is difficult to sort out the different contributing effects.

REFERENCES

1. F. Keffer, Handbuch der Physik, Springer-Verlag, New York, Vol. XVIII, part 2 (1966), p.1.
2. J.E. Rives, Transition Metal Chem. 7, 1 (1972).
3. L.J. de Jongh and A.R. Miedema, Adv. Phys. 23, 1 (1974). Recent theoretical developments are reviewed by J.F. Nagle and J.C. Bonner, Ann. Rev. Phys. Chem. 27, 291 (1976).
4. J. van den Handel, H.M. Gijsman, and N.J. Poulis, Physica 18, 862(1952).
5. K.W. Mess, E. Lagendijk, D.A. Curtis, and W.J. Huiskamp, Physica 34, 126 (1967).
6. T.A. Reichert and W.F. Giauque, J. Chem. Phys. 50, 4205 (1969); W.F. Giauque, R.A. Fisher, E.W. Hornung, and G.E. Brodale, J. Chem. Phys. 53, 3733 (1970).
7. R.B. Griffiths, Phys. Rev. Lett. 24, 715 (1970).
8. J.A. Griffin, S.E. Schnatterly, Y. Farge, M. Regis, and M.P. Fontana, Phys. Rev. B10, 1960 (1974); J.A. Griffin and S.E. Schnatterly, Phys. Rev. Lett. 33, 1576 (1974).
9. A.L.M. Bongaarts, Phys. Lett. 49A, 211 (1974); A.L.M. Bongaarts, Thesis, Eindhoven, 1975.
10. R.D. Spence and A.C. Botterman, Phys. Rev. B9, 2993 (1974).
11. (a) R.D. Spence, private communication.
(b) R.D. Spence, J. Chem. Phys. 62, 3659 (1975).
12. A.H. Morrish, "Physical Principles of Magnetism," J. Wiley and Sons, New York, 1965.
13. J. Smit and H.P.J. Wijn, "Ferrites," John Wiley and Sons, New York, 1959.
14. D.H. Martin, "Magnetism in Solids," M.I.T. Press, Cambridge, Mass., 1967.
15. T. Moriya, in "Magnetism," Edited by G.T. Rado and H. Suhl, Academic Press, New York, 1963, Vol. 1, Chapt. 3.
16. I. Dzyaloshinsky, J. Phys. Chem. Solids 4, 241 (1958).
17. T. Moriya, Phys. Rev. 117, 635 (1960); 120, 91 (1960).
18. A. Herweijer, W.J.M. de Jonge, A.C. Botterman, A.L.M. Bongaarts, and J.A. Cowen, Phys. Rev. B5, 4618 (1972); K. Kopinga, Q.A.G. van Vlimmeren, A.L.M. Bongaarts, and W.J.M. de Jonge, Physica, to be publ.
19. D.B. Losee, J.N. McElearney, G.E. Shankle, R.L. Carlin, P.J. Cresswell, and W.T. Robinson, Phys. Rev. B8, 2185 (1973).

20. J.P.A. Hijmans, W.J.M. de Jonge, P. van den Leeden, and M. Steenland, Physica 69, 76 (1973).
21. J.W. Metselaar and D. de Klerk, Physica 69, 499 (1973).
22. I.F. Silvera, J.H.M. Thornley, and M. Tinkham, Phys. Rev. 136, A695 (1964).
23. F. Keffer, Phys. Rev. 126, 896 (1962).
24. A.H. Cooke, K.A. Gehring, and R. Lazenby, Proc. Phys. Soc. 85, 967 (1965).
25. S. Merchant, J.N. McElearney, G.E. Shankle, and R.L. Carlin, Physica 78, 308 (1974).
26. R.E. Caputo, R.D. Willett, and J. Muir, Acta Cryst. B32, 2639 (1976).
27. H. Kobayashi, I. Tsujikawa, and S.A. Friedberg, J. Low Temp. Phys. 10, 621 (1973).
28. J.J. Pearson, Phys. Rev. 126, 901 (1962).

CHAPTER VIII

SELECTED EXAMPLES

A. INTRODUCTION

The discussion in this chapter concerns additional examples of interesting magnetic compounds that have been examined recently. It will be seen that all of the principles discussed earlier in this book come together here. No attempt is made at a complete literature survey, or even of all the work reported on a given compound. Consonant with the emphasis of this book will be the emphasis on results based on specific heat and magnetic susceptibility data, but especially we are trying to lay a firm foundation for a structural basis of magnetochemistry. The examples chosen are some of the best understood or most important ones, a knowledge of which should be had by everyone interested in this subject. We begin by briefly summarizing the single-ion properties of those ions which form compounds discussed below. Further information and references may be found in earlier sections of the book, as well as in Refs. 1-6.

B. SOME SINGLE ION PROPERTIES

1. Ti^{3+}, d^1, $^2T_2$

With but one unpaired electron, titanium(III) is a Kramers' ion and the ground state in a cubic field is $^2T_2$. This is therefore the prototype ion for all crystal field calculations, for the complication caused by electron-electron repulsions need not be considered. Despite this large body of theoretical work, the air-sensitive nature of many compounds of this ion has prevented a wide exploration of its magnetic behavior; more importantly, in those compounds where it has been studied, it rarely acts in a simple fashion or in any sensible agreement with theory.

The complete calculation of the g values and susceptibility of trivalent titanium have been outlined elsewhere (5,6), as well as in Chapt. III-D, above. A trigonal or tetragonal field is always present, which splits the $^2T_2$ state (in the absence of spin-orbit coupling) into an orbital singlet and doublet, separated by an amount δ. After adding spin-orbit coupling, Orton finds that for negative δ, $g_{\|} = g_{\perp} = 0$, the orbital and spin contributions cancelling. There appear to be no examples of this

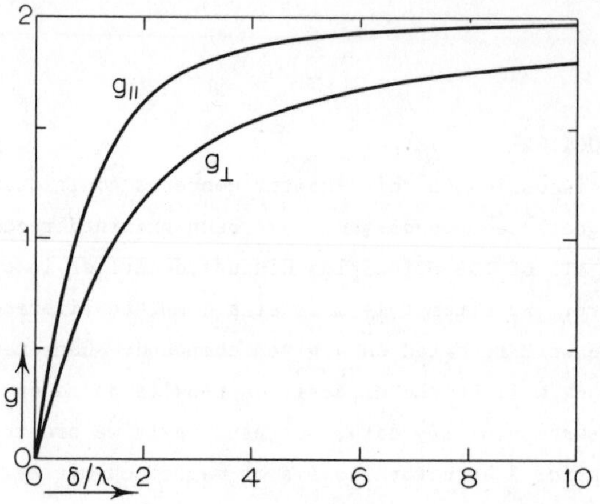

FIGURE 8.1 *Variation of $g_{\|}$ and $g_{\perp}$ as a function of tetragonal field parameter δ (for positive δ) for the $E\_$ doublet of the Ti^{3+} ion. From Ref. 5.*

situation, though $g_{\|}$ can be non-zero if the trigonal field is not very much weaker than the cubic one. Thus, $g_{\|} = 1.067$ and $g_{\perp} < 0.1$ for Ti^{3+} in Al_2O_3. In the situation of positive δ, the g values depend on the relative strengths of the axial field distortion (δ) and spin-orbit coupling (λ), as illustrated in Figure 8.1

Cubic cesium titanium alum, $CsTi(SO_4)_2 \cdot 12H_2O$, is probably the most thoroughly studied titanium(III) compound, largely because it is the least air-sensitive, but it is only recently that an understanding of its magnetic properties seems to be emerging. The initial impetus arose because it was assumed that its electronic structure would prove easy to determine, as well as attempts were made to discover its feasibility as a cooling salt for adiabatic demagnetization experiments (7). The salt is decidedly not ideal at low temperatures, however, and interest in the substance, at least, as a cooling salt, never developed very far.

Thus, it was assumed (1) that the $^2T_2$ (D) state was resolved by a static crystal field into three doublets separated by an energy of 100 cm^{-1} or so. At low temperatures, to first order, the system should then have acted as an effective spin-1/2 ion with a g of about 2 but interactions with the nearby excited states would cause some modification of the behavior. Measurements of the powder susceptibility (8) led to a g of only 1.12, and EPR on the diluted material (9) provided the values $g_{\parallel} = 1.25$, $g_{\perp} = 1.14$, and it proved impossible to rationalize these values with simple crystal field theory. It was suggested that the excited states in fact lay only some tens of cm^{-1} above the ground state, and the most recent measurements (10) of the magnetic susceptibility of $CsTi(SO_4)_2 \cdot 12H_2O$ suggested that in fact the lowest lying doublet was some 30 cm^{-1} above the ground state.

Over the years, Ti(III) has been studied by EPR in many alum lattices (11) although apparently the spectra have been assigned incorrectly to the presence of too many sites. The reason that Ti(III) does not behave in a straightforward fashion has been ascribed to a number of factors. Trigonal distortions have usually been assumed, due in part to size. After all, this is a relatively large ion with a radius estimated (12) as 0.67 Å, and many of the EPR experiments with titanium have been on isomorphous aluminum diluents, with the radius of Al(III) estimated as 0.53 Å. Size mismatch, followed with distortion of the lattice and a consequent effect on the electronic structure of the ion were thus frequently invoked. Indeed, a study (13) of Ti^{3+}: $[C(NH_2)_3]Al(SO_4)_2 \cdot 6H_2O$ suggested that the size mismatch was so important that titanium entered the guest lattice in effectively a random fashion, creating impurity centers as distinct from isomorphous replacement. Both signs of the trigonal field splitting parameter were observed within the one system. Static Jahn-Teller effect distortions have also been mentioned as a source of the problem.

Nevertheless, recent work suggests that the above arguments may not be correct, or at least not applicable to the pure, undiluted $CsTi(SO_4)_2 \cdot 12H_2O$. For one thing, the crystal structure analysis of the compound shows the structure to be highly regular and not distorted (14). Furthermore, important dynamic Jahn-Teller effects have been implicated as the likely source of the complexity of the problem.

In a recent series of papers (15-19) on both $CsTi(SO_4)_2 \cdot 12H_2O$ and $(CH_3NH_3)Ti(SO_4)_2 \cdot 12H_2O$, Walsh and his co-workers show that the energy level diagram of Ti^{3+} illustrated in Figure 8.2 accounts for all the observations

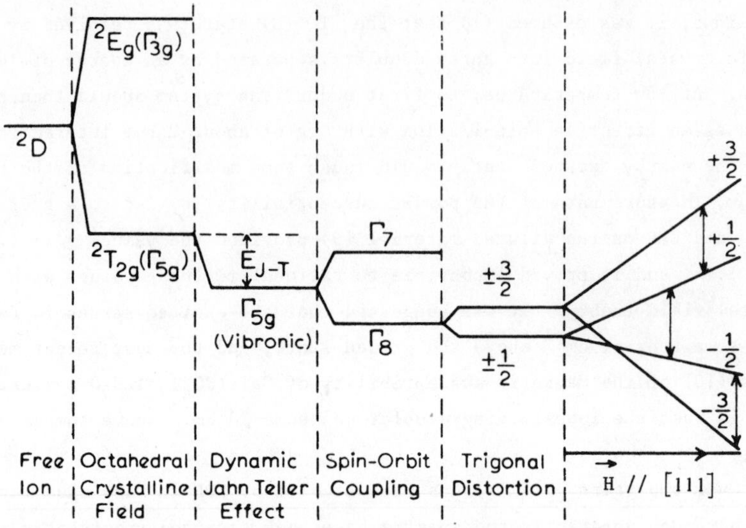

FIGURE 8.2 *Schematic energy-level diagram of* Ti^{3+}. *From Ref. 17.*

described above. In addition, spin-lattice relaxation studies on both compounds showed the presence at low temperatures of the Orbach process, which requires that there be energy levels accessible at rather low energies. In particular, for the Cs^+ compound, the Γ_{5g} electronic state is strongly coupled to the Γ_{3g} vibrational mode by means of a dynamic Jahn-Teller interaction. This fortuitously has the effect of canceling the major trigonal distortion, resulting in a quasi-Γ_8 quartet ground state, separated from Γ_7 by a mere 7 ± 1 cm^{-1}. A residual trigonal distortion gives rise to a zero-field splitting and three EPR lines, which disposes of the earlier model (11) of twelve magnetic complexes for Ti^{3+} in CsAl alum. The best values of the parameters on this model for Ti^{3+}: $CsAl(SO_4)_2 \cdot 12H_2O$ are (17), $g_{\parallel} = 1.1937 \pm 0.001$, $g_{\perp} = 0.6673 \pm 0.005$, and a zero-field splitting of some 0.002 K.

The resolution of this long-standing problem illustrates the fact that, at least for trivalent titanium, static crystal field calculations such as those mentioned earlier are useful pedagogical devices, but bear little relationship to the true nature of the problem.

Optical spectra (20), EPR spectra (21), as well as spin-lattice relaxation studies (22), suggest that Ti(III) is severely distorted in the acetylacetonate. This molecule, with $g_{\parallel} = 2.000$, $g_{\perp} = 1.921$, is also unusual in that EPR spectra may be observed even at room temperature. A complete theory of Ti(III) in trigonal environments has been published (23), but serious discrepancies with experimental results for several systems were found. The problem probably lies with the fact that the Jahn-Teller effect was ignored, and that attention was centered on the g-values obtained by EPR. Since these are characteristic of the lowest-lying or ground state, the contributions of the (mostly unknown) low-lying states was not considered.

We have paid especial attention to this ion only because one can conclude that the single-ion properties of titanium(III) are not at all simple, and caution must be applied in the study of its compounds. The T_c of $CsTi(SO_4)_2 \cdot 12H_2O$ has not been reported, but should be quite low. No other T_c of a simple, insulating compound of Ti(III) has been reported either.

2. V^{3+}, d^2, $^3A_2$

This ion is of more interest for its single-ion properties than for its magnetically-ordered compounds, the latter of which there are hardly any. The zero-field splitting is large, being of the order 5-15 K in the compounds examined to date. The large splitting, which is easily measured from either susceptibilities or specific heats, arises because the octahedral ground state is in fact a ^3F; the $^3A_2(F)$ state becomes the ground state because of axial distortions, but the excited $^3E(F)$ component lies generally only about 1000 cm^{-1} above the $^3A_2$. What is of especial interest is that of the half-dozen cases in which the zero-field splitting of V^{3+} has been evaluated, it is always positive (doublet above the singlet). There does not seem to be any theoretical rationale for this fact, nor does it seem to be required by the theory available (24).

One consequence of the large zero-field splittings in vanadium is that the allowed $\Delta m = \pm 1$ paramagnetic resonance transition has only been observed in V^{3+}: Al_2O_3, by going to very high fields with a pulsed magnetic field (25); the low-lying orbital levels also give rise to short spin-lattice relaxation times which require that helium temperatures be used for EPR. The forbidden $\Delta m = \pm 2$ transition, however, has been observed with several samples, such as with V^{3+}: $[C(NH_2)_3]Al(SO_4)_2 \cdot 6H_2O$ (13) and V^{3+}: Al_2O_3 (26). The g-values for trigonally-distorted V^{3+} are typically $g_{\parallel} \approx$ 1.9 and $g_{\perp} \approx 1.7$-1.9.

With the exception of some poorly-understood oxides, there are no well-characterized examples of magnetically-ordered vanadium(III) compounds. But, as was discussed in Chapt. V-G with $Cu(NO_3)_2 \cdot 2\frac{1}{2}H_2O$ as example, a field-induced magnetic ordering may be able to be caused to occur with vanadium (and nickel(II), as well). The crystallographic symmetry restrictions are greater here, however, than with the copper nitrate system. There, the excited spin-triplet occurred because of isotropic exchange; anisotropic contributions are quite small in that system. When the excited doublet arises, as with V and Ni salts, because of zero-field splittings, the splitting in a field is quite anisotropic. This was illustrated in Chapt. III with the calculation of the energy levels of trigonal nickel. The strict symmetry requirement is then that all the molecules in the unit cell lie such that an external magnetic field be able to be applied parallel to all the z axes of the constituent metal ions. Several such systems have recently been discovered (27) ($NiSnCl_6 \cdot 6H_2O$; $C(NH_2)_3V(SO_4)_2 \cdot 6H_2O$; $Cs_3VCl_6 \cdot 3H_2O$; $[Ni(C_5H_5NO)_6](ClO_4)_2$) and are currently under active investigation.

3. VO^{2+}; d^1, $^2B_2$

This ion is interesting because of its binuclear nature. Its electronic configuration allows paramagnetic resonance to be observed easily, and many studies have been reported. These include the ion as diluent in a variety of crystals, and even when dissolved in liquids at ambient temperatures. The g values are isotropic at about 1.99, as anticipated, and simple, spin-only magnetism is frequently observed. Molecular structures are usually quite distorted.

No magnetically-ordered vanadyl compounds have yet been reported but more work in this area should be fruitful. Chain magnetism has been suggested for some vanadyl acetates, and a number of dimers have been reported to behave magnetically somewhat like copper acetate (28).

4. Cr^{3+}, d^3, $^4A_2$

Since chromium(III) is a spin-3/2 ion, zero-field splittings are usually found in its compounds. Since the spin-orbit coupling constant is relatively small, however, the zero-field splittings observed to date are also quite small. Thus, EPR measurements lead to fairly typical values of only 0.592 cm^{-1} for D in Cr^{3+}:Al $acac_3$ and 0.00495 cm^{-1} for Cr^{3+} in Co en_3 $Cl_3 \cdot NaCl \cdot 6H_2O$. The g value is almost always isotropic at about 1.98 or 1.99. One would therefore expect chromium to furnish relatively good examples of Heisenberg magnets, yet there are few such studies. One reason probably lies with the kinetic inertness of chromium(III), which hinders the synthesis of appropriate materials; another reason lies with the fact that exchange interactions with chromium(III) are so weak that most of its compounds that are known to undergo long-range order do so at temperatures well below 1 K. A compensating factor, however, is that chromium(III) alums, in particular, have found wide use as magnetic thermometers and as cooling salts for adiabatic demagnetization (29).

5. Mn^{2+}, d^5, ^6S

The manganese ion is the source of a vast portion of the literature on magnetism. The reasons for this are straightforward. Chemical problems of synthesis are usually minimal. Since it has an odd number of electrons, it is a Kramers ion, and EPR spectra may be observed under a wide variety of conditions. The g values are isotropic at about 2.0 and so manganese is the best-known example of a Heisenberg ion. Zero-field splittings are usually small - of the order of 10^{-2} cm^{-1}. On the other hand, several systems will be mentioned later which appear to have quite large zero-field splittings, and the large value of the spin sometimes causes dipole-dipole interactions to be important.

Strong-field or spin-paired manganese has been studied by EPR in such compounds as $K_4Mn(CN)_6 \cdot 3H_2O$. Spin-orbit coupling effects are important, g-value anisotropy is found, and the optical spectra are dominated by charge transfer effects. Relatively few magnetic studies pertain to this electronic configuration of the ion.

6. Fe^{3+}, d^5, ^6S

The simple compounds of iron(III) behave quite simply, but the complicated compounds behave in a very complicated fashion! Three different electronic ground states are known for this ion, depending on the local symmetry, and in fact we shall discuss below a series of compounds in which an equilibrium between two of those states is found.

The Fe(III) ion (and also isoelectronic Mn(II) has been studied extensively by EPR. In a weak crystal field, the ^6S state of d^5 has no nearby crystal-field states, and thus the EPR is easily detected over a large range of temperatures in any crystal-field symmetry. Furthermore, since this is an odd-electron (Kramers) ion, resonances are always detected even with large zero-field splittings. The g values are isotropic at about 2.0. In addition to the usual D and E zero-field terms, higher-order terms (denoted a and F) are usually required in the analysis of the EPR spectra of iron and manganese, but these are too small to affect their other magnetic properties. The parameter D is usually substantially less than 1 cm^{-1} for iron(III), but the EPR linewidths are often quite broad.

Since manganese(II) is such a good example of a Heisenberg ion, one would anticipate that iron(III) would also provide many such good examples. In fact, as with chromium(III), there are relatively few examples reported to date of cooperative magnetic interactions with spin-free iron(III), although a number of clusters have been studied.

Tetrahedral iron(III) is expected to behave similarly, but there are few data of relevance for our present purposes. Five-coordinate iron(III) is quite unusual, and is therefore dealt with separately in Section C-6. Iron(III) in strong crystalline fields has but one unpaired electron and a $^2T_2$ ground state. Three orbital states are then low-lying, and spin-orbit coupling effects become very important. Low temperatures are required to detect the EPR, and the g values vary greatly from 2. One well-known antiferromagnet with this electronic configuration is $K_3Fe(CN)_6$ (30).

7. Fe^{2+} (d^6) and Cr^{2+} (d^4)

These ions are grouped together because they are relatively unfamiliar but have essentially similar electronic states. A few examples of magnetic interactions in chromous compounds have recently appeared, but ferrous compounds that display magnetic ordering phenomena are well-known.

The problem with both these spin-2 ions lies not only with their air-sensitive nature, but especially with the large number of electronic states that they exhibit. Being non-Kramers ions, there are relatively few EPR studies: short spin-lattice relaxation times are found, and large zero-field splittings. For example, D has been reported as 20.6 cm^{-1} for Fe^{2+}: $ZnSiF_6 \cdot 6H_2O$; $FeSiF_6 \cdot 6H_2O$ was discussed above in Chapt. III-G. The g-values deviate appreciably from 2.

Magnetic ordering is known with such materials as $FeCl_2.4H_2O$, ferrous oxalate, $FeCl_2.2H_2O$, and ferrous formate. Mössbauer (4-340 K) and magnetic susceptibility (80-310 K) studies have recently been reported on a number of $[FeL_6](ClO_4)$ compounds, where L is a sulfoxide or pyridine N-oxide (31). Spin-paired ferrous ion is of course diamagnetic; a large number of studies have been reported concerning an equilibrium between the two spin-states (32).

8. Co^{2+}, d^7

The ground state of this ion will be explored in detail, because cobalt illustrates so many of the concepts important in magnetism, as well as because it provides so many good examples of interesting magnetic phenomena. In particular, cobalt(II) is the best Ising ion, the only XY ion to date, and as will be described below, under the right conditions it could even be a Heisenberg ion!

Cobalt(II) with three unpaired electrons, exhibits an important orbital contribution at high temperatures, and a variety of diagnostic rules have been developed to take advantage of this behavior (33). The lowest electronic states in octahedral fields are illustrated in Fig. 3.14, where it will be seen that the effects of spin-orbit coupling (λ_0 = 180 cm^{-1}) and crystalline distortions combine to give a spin-doublet ground state, separated by 100 cm^{-1} or so from the next nearest components. This ground state is the interesting one in cobalt magnetochemistry, so we restrict the remainder of the discussion to low temperatures, at which the population of the other states is small (34).

The doubly-degenerate level is an effective spin-1/2 state, and so unusual features may be anticipated. The theory is essentially due to Abragam and Pryce (32) and Kambe (36), and has been reviewed by Bates and Wood (34).

Two parameters have been introduced that are useful for the empirical representation of magnetic data. The first of these, a Landé factor usually called α (35) refers to the strong-field (α = 1) and weak-field (α = 3/2) limits and its diminution in value from 3/2 is a measure of the orbital mixing of $^4T_1(F)$ and $^4T_1(P)$. The lowest electronic level in an axial field with spin-orbit coupling is a Kramers doublet and so cannot be split except by magnetic fields. The orbital contribution of the nearby components of the $^4T_1(F)$ state causes the ground doublet in the weak cubic field limit to have an isotropic g = 4.33, a result in agreement with experiments on cobalt in MgO (37), but large anisotropy in the g value is expected as the crystal field becomes more distorted. The three orthogonal g values are expected to sum in first order to the value of 13 (35, 38).

The second parameter, δ, is an axial crystal field splitting parameter that measures the resolution of the degeneracy of the $^4T_1$ state, and thus is necessarily zero in a cubic crystal. The isotropic g value is then $(2/3)(5+\alpha)$, to first order (37). In the limit of large distortions, δ may take on the values of $+\infty$ or $-\infty$, with the following limiting g values resulting:

$$\delta = +\infty : g_{\parallel} = 2(3 + \alpha), \; g_{\perp} = 0$$
$$\delta = -\infty : g_{\parallel} = 2, \; g_{\perp} = 4$$

For a given α, the two g values are therefore functions of the single parameter δ/λ and so they bear a functional relationship to each other. Abragam and Pryce have presented the general result (35), but there are more parameters in the theory than can usually be obtained from the available experimental results. With the approximation of isotropic spin-orbit coupling, they derive (to first order) the following equations which provide a useful estimate of crystal distortions,

$$g_{\parallel} = 2 + 4(\alpha+2)\,[\,(3/x^2 - 4/(x+2)^2)/(1 + 6/x^2 + 8/(x+2)^2)] \qquad (8.1)$$

$$g_{\perp} = 4\,[\,(1 + 2\alpha/(x+2) + 12/x(x+2))/(1 + 6/x^2 + 8/(x+2)^2)] \qquad (8.2)$$

where x is a dummy parameter which for the lowest energy level is positive with limiting values of 2 (cubic field, $\delta=0$), 0 ($\delta=+\infty$), and ∞ ($\delta=-\infty$). The splitting parameter δ is found as

$$\delta = \alpha\lambda\,[\,(x+3)/2 - 3/x - 4/(x+2)] \qquad (8.3)$$

The observed g values for octahedral environments with either trigonal or tetragonal fields should therefore all lie on a universal curve. Such curves are illustrated by Abragam and Pryce (35) and Orton (5), and reproduced in Figure 8.3, and a satisfactory relationship of theory and experiment has been observed. The observation of $g_{\parallel}$ and $g_{\perp}$ allows the solution of Eqs. (8.1) and (8.2) for α and x, and these parameters in turn may be applied to Eq. (8.3). Cobalt tends to act in Ising fashion when $g_{\parallel} >> g_{\perp}$, and as an XY ion when $g_{\perp} >> g_{\parallel}$. Exchange constants determined for cobalt as an $S'=1/2$ ion are four times larger than those calculated when the true spin of 3/2 is used.

The situation is far different when the cobalt(II) ion resides in a tetrahedral geometry. A $^4A_2$ ground state results, and with a true spin of 3/2, the g values are slightly anisotropic and lie in the range of 2.2 to 2.4. A zero-field splitting of the spin-quartet is of course probable, but if the exchange constant should be large with respect to this splitting, then one expects Heisenberg magnetic behavior. There are as yet no data that provide an example of this situation, primarily because the zero-field splitting is generally quite large for this ion.

Thus, for 2D large and positive, the $\pm$ 1/2 state lies low, and if J is small with respect to 2D, one expects to have an effective spin-1/2 system with, at least for $g\mu_B H << D$, $g_{\parallel}' = g_{\parallel}$, and $g_{\perp}' = 2g_{\perp}$. One therefore finds that $J_{\perp}/J_{\parallel} \approx (g'_{\perp}/g'_{\parallel})^2 \approx 4$ and anticipates that the system will behave more or less as an XY magnet. The compounds Cs_2CoCl_4 and $(Et_4N)_2CoCl_4$ in fact have zero-field splittings of 13 to 15 K, and Cs_2CoCl_4 has been shown to behave like an XY magnet.

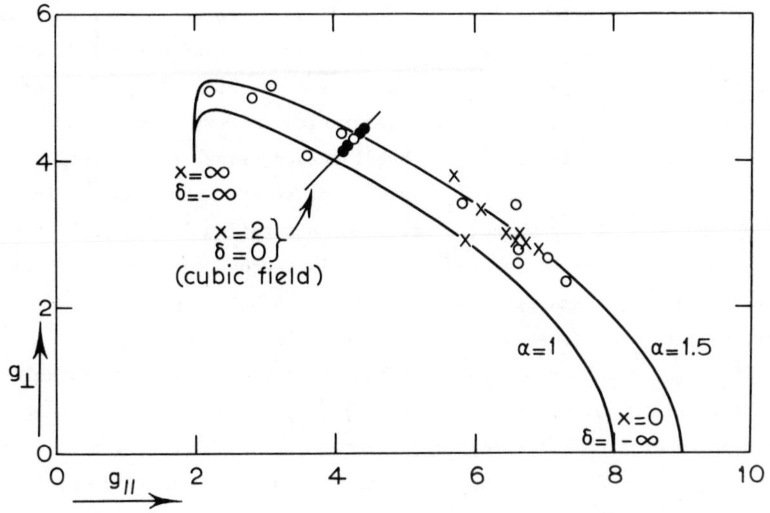

FIGURE 8.3 *Experimental and theoretical g-values for the configuration $t_2^5e^2$. Where the site symmetry is rhombic $g_{\perp}$ is taken as the mean of g_x and g_y. From Ref. 5.*

On the other hand, if 2D is large and negative, the ± 3/2 state lies lower. The system again acts as an effective spin-1/2 system, but in this case $g'_{\parallel} = 3g_{\parallel}$ and $g'_{\perp} = 0$. Ising magnetism is then implied, and has been found in the A_3CoX_5 (A = Cs, Rb; X = Cl, Br) series of compounds, where 2D is again of the order of 15 K, but of negative sign.

It is fascinating that the entire interplay of single-ion anisotropy effects and magnetic ordering models can be illustrated by this one ion.

9. Ni^{2+}, d^8, $^3A_2$

The electronic structure of this ion has been dealt with ad nauseum in the previous portions of this book. The reasons for this should be clear: nickel forms complexes with a wide variety of ligands, as well as the fact that several stereochemistries are common. Restricting the discussion to octahedral complexes, the spin of this ion is large enough to make it a sensitive probe of a variety of phenomena, and yet it is not too high to prevent theoretical analyses. Spin-orbit coupling is important enough to cause the g values to deviate measurably from the free ion values, to values typically in the neighborhood of 2.25. Yet, the g values are commonly found to be isotropic or nearly so, and spin-orbit coupling causes no other problems. The lowest-lying excited states are far enough away (usually at least 8000 cm^{-1}) so as to be unimportant magnetically. Paramagnetic resonance is generally observed without difficulty, frequently at room temperature; on the other hand, nickel(II) is a non-Kramers ion, and the resolution of degeneracies thereby allowed is occasionally observed. Thus, as will be discussed below, the zero-field splitting in $[Ni(C_5H_5NO)_6](ClO_4)_2$, as well as several other salts, is large enough so that paramagnetic resonance absorption is prevented at x-band, even at helium temperatures. This is because the energy of the microwave quantum is smaller than the separation between the $|0\rangle$ and $|-1\rangle$ states, the states between which a transition is allowed.

The dominant feature of the magnetochemistry of octahedral nickel(II) is the zero-field splitting of the ground, $^3A_2$, state. The consequences of this phenomenon have been explored above, extensively. Neglecting the rhombic (E) term for the moment, the parameter D/k is usually found to be of the order of a few kelvins. That puts important magnetic anisotropy and Schottky anomalies in the specific heat at a convenient temperature region, the easily-accessible helium one. What is of interest is that both signs of D are found, being for example, positive in $Ni(NO_3)_2 \cdot 6H_2O$ (39), $NiSnCl_6 \cdot 6H_2O$ (40), and $[Ni(C_5H_5NO)_6](ClO_4)_2$ (41) and negative in $NiCl_2 \cdot 4H_2O$ (42), $NiCl_2 \cdot 2py$ (43) and $NiZrF_6 \cdot 6H_2O$ (40). The magnitude of D varies widely, being as small as +0.58 K in $NiSnCl_6 \cdot 6H_2O$ and as large as +6.26 K in $[Ni(C_5H_5NO)_6](ClO_4)_2$, and even -30 K in the linear chain series $NiX_2 \cdot 2L$, X = Cl, Br; L = pyridine, pyrazole (43).

The sign and magnitude of D has important consequences for nickel, especially with regard to magnetic ordering phenomena. When $|D|$ (or $|D| + |E|$) is much smaller than the exchange constant J, then magnetic interactions will occur at higher temperatures, and the sign of D is unimportant. But, when $|D| > |J|$, several new situations occur. Thus, if D is negative, as the temperature decreases the spin-singlet state will be depopulated, leaving a doubly-degenerate ground state. A compound with nickel in this situation will order at some low temperature, but as an effective spin-1/2 ion. The anisotropic g values ($g'_{\parallel} = 2g_{\parallel} \approx 4.5$, $g'_{\perp} = 0$) then cause nickel to be an Ising ion! This situation has been observed in $NiBr_2.2pz$ (43).

When the zero-field splitting is positive and larger than the exchange interaction, magnetic ordering need not occur. That is, as the temperature is lowered and the spin-doublet is depopulated, the remaining ground state has no degeneracy at 0 K. In other words, all the magnetic entropy can be removed by cooling alone, and long-range magnetic ordering is not required in order to find zero entropy at 0 K. Moriya (44) shows, in a molecular field approach, that the condition for a critical (Néel) temperature is $|D-E| < 2zJ$.

Even though long-range order therefore won't necessarily occur, exchange interactions may still be present and even observable. Such exchange interactions have been termed subcritical, and been observed in such salts as $Ni(NO_3)_2 \cdot 6H_2O$ (39) and $[Ni(C_5H_5NO)_6](ClO_4)_2$ (41). The presence of such subcritical interactions is revealed through their contribution to the effective magnetic field at a nickel ion when an external field is applied. Therefore, while both the low field susceptibility and zero-field specific heat of the system would appear at first glance to be those of an assembly of noninteracting ions, the single-ion parameters needed to fit the data would be the same only if the interactions were included in the calculation of the susceptibility.

Since nickel(II) has the same ground state as vanadium(III), field-induced magnetic ordering as described above in Section 2 may also be found with nickel. The first example of this phenomenon has been found with $[Ni(C_5H_5NO)_6](ClO_4)_2$, and will be discussed in Section C-8. The magnetic properties of tetrahedral nickel have been well-characterized at high temperatures (3) but are of little interest at low temperatures. As referred to in Chapt. III-F, tetrahedral nickel is expected to be diamagnetic in the helium region. Although many compounds have been prepared with planar, four-coordinate nickel, they are diamagnetic at all temperatures.

10. Cu^{2+}, d^9, ^2E

The single-ion magnetic properties of copper(II) are fairly straightforward. Spin-orbit coupling is large, causing the g values to lie in the range 2.0 to 2.3, but because copper has an electronic spin of only 1/2, there are no zero-field splitting effects. The g values are often slightly anisotropic, being for example 2.223 (∥) and 2.051 (⊥) in $Cu(NH_3)_4SO_4 \cdot H_2O$. Nevertheless, copper compounds usually provide good examples of Heisenberg magnets.

The one problem that does arise with this ion is that it rarely occupies a site of high symmetry; in octahedral complexes, two trans ligands are frequently found substantially further from the metal than the remaining four. This has led to many investigations of copper as the Jahn-Teller-susceptible ion, par excellance. Dynamic Jahn-Teller effects have also been frequently reported in EPR investigations of copper compounds at low temperatures.

11. Lanthanides

Some of the properties of the lanthanide ions were described in Chapter I. Because of the importance of spin-orbit coupling, it is difficult to generalize about the properties of a given ion; the magnetic behavior of a lanthanide ion varies widely, depending on the particular compound it is in. Most of the low-lying energy levels are mixed by both spin-orbit coupling and the crystalline field, and so the ground state can rarely be characterized by a single value of J_z. The g values are generally quite anisotropic and deviate substantially from 2. EPR usually can be observed only at temperatures of 20 K or lower. Furthermore, several of the ions are non-Kramers ions, Eu^{3+} ($^7F_0$) being an example. With a J = 0 ground state, no EPR can be observed. For these reasons, as well as the fact that there are relatively few magnetic studies of compounds of the sort that this book is concerned with, we describe very briefly some of the properties of only two ions.

The first, gadolinium(III), is fairly straightforward. With a half-filled shell of seven unpaired electrons, the ground state is ^8S, the effect of the crystalline field is small, the zero-field splittings are generally very small, and the long spin-lattice relaxation times usually allow EPR to be observed at room temperatures. The EPR spectrum resembles to a certain extent, that of manganese(II), but additional terms are required because of the high value of the spin. The g values are isotropic at about 2. Salts such as $Gd_2(SO_4)_3{\cdot}8H_2O$ have long been investigated for adiabatic demagnetization purposes because of the high spin of the ion.

The situation with cerium(III) (and most of the rest of the lanthanide ions) is substantially different. With the f^1 electronic configuration, a $^2F_{5/2}$ state lies lowest for this Kramers ion with the $^2F_{7/2}$ state some 2000-2500 cm^{-1} higher. The degeneracy of the three components of the ground state ($J_z = \pm 1/2, \pm 3/2, \pm 5/2$) is generally resolved, and the g values of each level differ. Since the splitting between the levels is not large, there is also a mixing of the different states. The details of the calculation have been summarized by Orton (5), so we merely illustrate the situation by pointing out that a first-order calculation of the g values in a tetragonal field yields $g_{\parallel} = 6/7$, $g_{\perp} = 18/7$ for the ± 1/2 state; $g_{\parallel} = 18/7$, $g_{\perp} = 0$ for $J_z = \pm 3/2$, and $g_{\parallel} = 30/7$, $g_{\perp} = 0$ for the ± 5/2 state.

The calculation, and resulting g values, differs if the local symmetry is changed. Orton (5) illustrates this case with $Ce(EtOSO_3)_3 \cdot 9H_2O$, in which the cerium has a three-fold symmetry axis together with a perpendicular reflection plane. Experimentally (45), when the cerium is diluted by the isomorphous La(III) salt, resonance is observed from two doublets separated by only 3 cm^{-1}, the g factors being

Lower doublet state	$g_{\parallel} = 0.955$
	$g_{\perp} = 2.185$
Upper doublet state	$g_{\parallel} = 3.72$
	$g_{\perp} = 0.2$

which are in approximate agreement with theoretical values for being primarily the ± 1/2 and ± 5/2 doublets, respectively. In the pure or undiluted salt, magnetic susceptibility measurements show that the order of the two levels is inverted; the ± 5/2 state is also lower in $Ce{:}LaCl_3$, where the g values observed by EPR are $g_{\parallel} = 4.0366$ and $g_{\perp} = 0.17$ (46). It should be apparent from this example alone that the magnetochemistry of the lanthanides is more difficult to explore than that of the 3d ions.

We conclude by mentioning CMN, $Ce_2Mg_3(NO_3)_{12} \cdot 24H_2O$, which remains today as an important magnetic thermometer (29). The salt is very anisotropic, with $g_{\parallel} \leq 0.026$ and $g_{\perp} = 1.840$. The ± 1/2 state lies lowest, with the next state, ± 3/2, at 25.2 cm^{-1} above. Since the metal ions are so dilute, the salt obeys the Curie-Weiss law with a very small θ down to temperatures below 0.01 K, and dipole-dipole interaction alone is able to account for the low-temperature magnetic and thermal properties. The presence of the first excited state at only 25 cm^{-1} above the ground state leads to an exponential temperature dependence of the spin-lattice relaxation time, and in fact it was work on this salt that led to the discovery of the Orbach relaxation process (Chapt. II-F).

C. SOME EXAMPLES

1. Iron(III) Methylammonium Sulphate

This compound, $(CH_3NH_3)Fe(SO_4)_2 \cdot 12H_2O$, provides an example of the utility of specific heat measurements of magnetic systems. In particular,

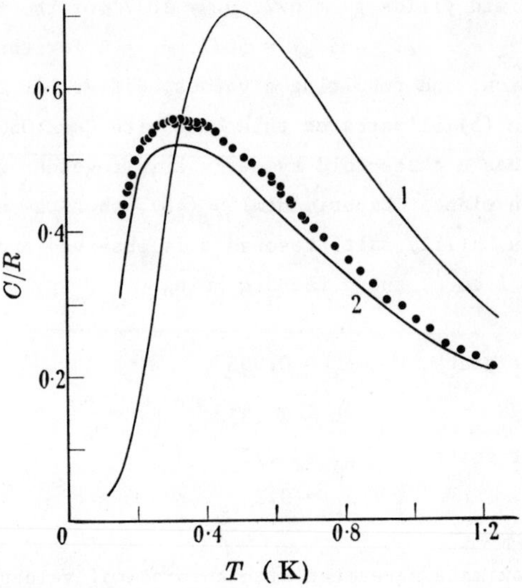

FIGURE 8.4 *Specific heat at temperatures between 0.17 and 1.2 K. •, experimental points; curve 1, calculated Stark specific heat, assuming ± 5/2 level lowest; curve 2, calculated Stark specific heat, assuming ± 1/2 level lowest. From Ref. 47.*

a Schottky specific heat was observed (47) and allowed an unambiguous assignment of the energy levels of the ground state.

The alums, of which this compound is an example, have a cubic crystal structure and contain hexaquometal(III) ions. The long-standing interest in these compounds arises from the fact that they have low ordering temperatures, and thus have been explored extensively below 1 K as possible magnetic cooling salts and magnetic thermometers. In the current example, it had been shown by EPR that the ^6S ground state of the iron atom was split into the three doublets, $|\pm 1/2\rangle$, $|\pm 3/2\rangle$, and $|\pm 5/2\rangle$, with successive splittings of 0.40 and 0.73 cm^{-1}. The power of EPR lies with the ease and accuracy that zero-field splittings of this magnitude can be obtained; the fault of EPR lies with the fact that the sign of the zero-field splitting cannot be directly determined, and so it was not known which state, $|\pm 1/2\rangle$ or $|\pm 5/2\rangle$, was the ground state of the system.

The calculation of the anticipated Schottky behavior is simple for a three-level system, for the partition function is merely

$$Z = 1 + \exp(-\delta_1/kT) + \exp(-\delta_2/kT), \tag{8.4}$$

where δ_1 and δ_2 may be either both positive or both negative. The measured specific heat is illustrated in Figure 8.4, along with the specific heat calculated from Eq. (8.4) for both signs of δ_1 and δ_2 Curve 1 corresponds to the $|\pm 5/2\rangle$ state low and curve 2, which agrees quite well with experiment, corresponds to the $|\pm 1/2\rangle$ state as the ground state.

2. $CaCu(OAc)_4 \cdot 6H_2O$

This compound has had a tortuous, recent history in magnetic studies. The problem with it is that it is an interesting compound from a structural point of view. This has led several investigators to postulate the presence of measureable magnetic interactions which other investigations have never been able to verify.

Thus, a portion of the tetragonal structure (48) is illustrated in Figure 8.5. The structure consists of polymeric chains of alternating

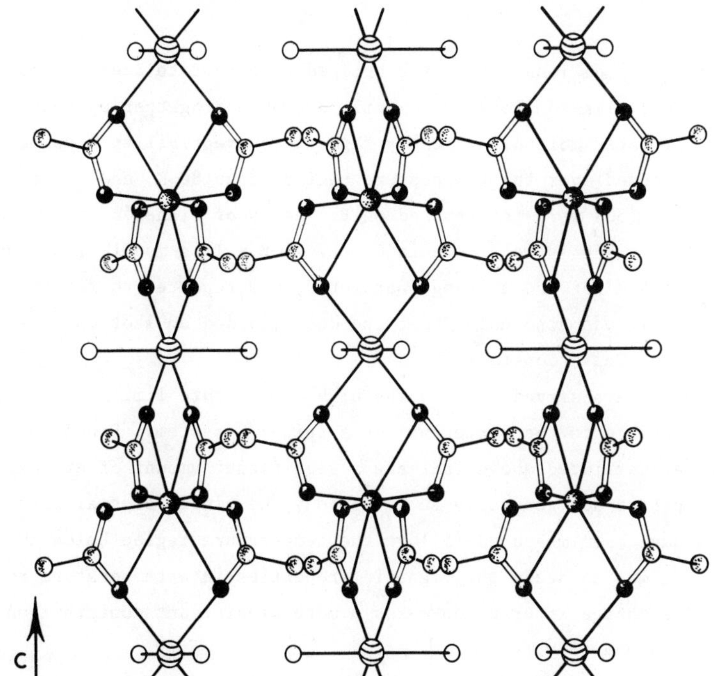

FIGURE 8.5 *Portion of a polymeric chain of* $CaCu(OAc)_4 \cdot 6H_2O$. *After Ref. 48.*

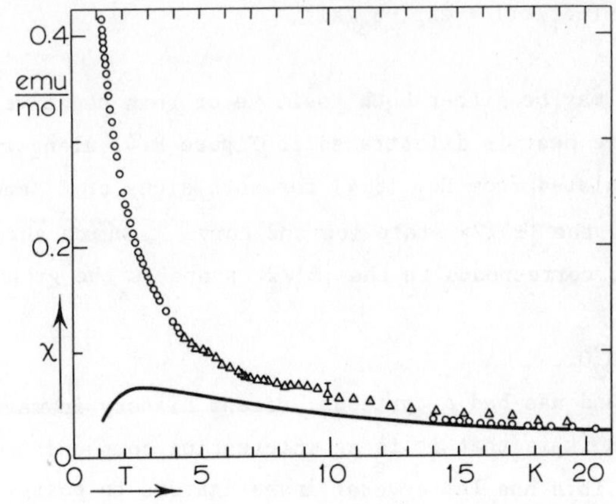

FIGURE 8.6 *Experimental parallel susceptibility of $CaCu(OAc)_4 \cdot 6H_2O$ compared to the theoretical prediction for a linear chain with J/k = -1.4 K. From Ref. 51.*

copper and calcium ions, which are bridged by bidentate acetate groups. These chains, aligned along the c axis, are bound together by solvent cages of 12 water molecules. Early magnetic susceptibility measurements (49,50) extended over the temperature region above 80 K, and the results of one study (50) were interpreted on the basis of Fisher's linear chain Ising model, which is mentioned in Chapt. VI-B. This model with a value of J/k = -1.4 K for the exchange parameter, was reported to fit the data very well. It was also noted that the data yielded an isotropic value of -10 K for the Weiss constant.

It will be observed that in the high temperature limit, Eqs. (6.4) and (6.5) reduce to the Curie law for a spin-1/2 system. Since both of the parameters listed above indicate a significant amount of exchange, and since copper is not generally an Ising ion, the single crystal susceptibilities have been measured (51) in the temperature region below 20 K. The purpose was to study the magnetic properties in a temperature region where the exchange interaction makes a more significant contribution to the measured quantity.

In Figure 8.6, we show the behavior calculated from Fisher's equation in the low temperature region with the exchange parameter J/k = -1.4 K. Only the parallel susceptibility is illustrated, but the following conclusions also apply to the perpendicular direction. Two sets of data measurements (51) through the temperature region 1-20 K are illustrated, and it will be seen that experiment does not agree in any fashion with the calculated behavior. That is, Curie-Weiss behavior with a θ of only -0.03 K is observed (51-53) and thus only a weak exchange interaction is in fact present.

Line-width studies by EPR were also interpreted (54) in terms of a one-dimensional magnetic interaction, but the analysis is invalidated because a large exchange constant was used (52). Specific heat studies below 1 K have also provided no evidence for significant magnetic interactions (53), but single crystal susceptibilities at very low temperatures have finally clarified the situation (55). The susceptibilities parallel and perpendicular to the principal axis exhibit broad maxima at about 40 mK. The data were successfully fit, not to the linear chain model, but to a nearly-Heisenberg two-dimensional antiferromagnet! Only the slightest anisotropy was observed, with $J_{c_\parallel}/k = -21.4$ mK and $J_{c_\perp}/k = -22.5$ mK. Long-range order has not been observed above 27 mK. The very weak exchange interactions are consistent with the geometrical isolation of the copper ions from one another.

A final note concerns the specific heat of $CaCu(OAc)_4 \cdot 6H_2O$ in the helium region (53,56) which exhibits a broad, Schottky-like maximum that evidently, from the discussion above, must be non-magnetic in origin. Reasoning by analogy to the reported (57) behavior of some hexammine nickel salts, which provide specific heat maxima which are due to hindered rotation of the NH_3 groups about the M-NH_3 axis, this maximum was analyzed in terms of hindered rotation of the methyl groups of the acetate ligand. Substitution of deuterium on the methyl group gave consistent results, and illustrates nicely the fact that all phenomena observed with magnetic compounds at low temperature need not be magnetic in origin.

3. Hydrated Nickel Halides

The hydrated halides of the iron-series ions are among the salts most thoroughly investigated at low temperatures. For example, $MnCl_2 \cdot 4H_2O$ with $T_c = 1.62$ K is perhaps the most famous antiferromagnet; a brief, now out-dated list of references to important work on this salt amounts to a dozen items (58). Fame is a matter of taste, however, and some would argue that $CuCl_2 \cdot 2H_2O$, $T_c = 4.3$ K is the most important antiferromagnet (59). While there are other important candidates for the position as well, these two salts are of the class described. Since it would be difficult to discuss all of these properly, and since the hydrated halides of nickel have continued to be of interest and also offer some nice pedagogical examples, the discussion here will be limited to that class of salts. The limitation will be seen to be hardly restrictive.

We begin with the chemical phase diagram (60) of nickel chloride in water, which is illustrated in Figure 8.7. The search for new magnetic materials depends heavily on the application of the information contained

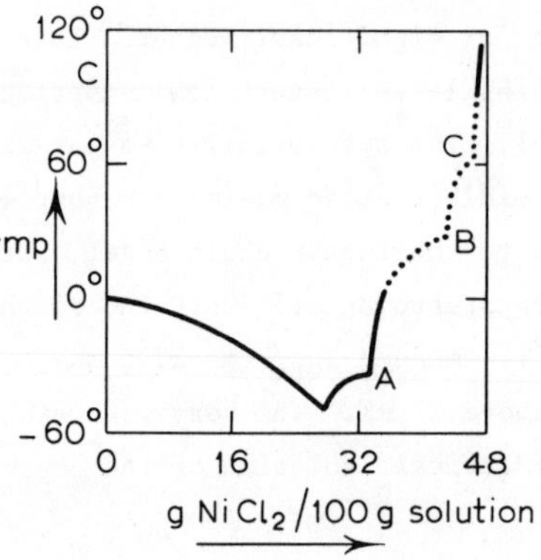

FIGURE 8.7
Phase diagram of $NiCl_2$-H_2O system. Point A corresponds to -33°C, B to 28°C, and C to 64°C. From Ref. 60.

in such diagrams, and a careful exploration of such phase studies can be quite fruitful. In this simple example, four hydrates of nickel chloride will be seen to exist, although the heptahydrate, which must be obtained below -33°C, is little known. The hexahydrate, $NiCl_2 \cdot 6H_2O$ is the best known salt in this series and may be obtained from an aqueous solution over the wide temperature interval of -33 to +28°C. The tetrahydrate is found between 28 and 64°C, and the dihydrate above 64°C; though less well known than the hexahydrate, both of these salts have been studied at low temperatures recently, and are of some interest. The critical temperatures for each of these salts is given in Table 8.1.

TABLE 8.1 Critical Temperatures of Hydrated Nickel Chlorides.

	T_c,K	Reference
$NiCl_2 \cdot 6H_2O$	5.34	61
$NiCl_2 \cdot 4H_2O$	2.99	62
$NiCl_2 \cdot 2H_2O$	7.258	63
$NiCl_2$	52	64

The T_c for $NiCl_2$ has been included in Table 8.1 for comparison, because it shows the effect of dilution on superexchange interactions. Without any knowledge of the structures of the materials, one would guess that the metal ions will be further apart in the hydrates than in the anhydrous material; furthermore, since chloride is more polarizable than water, one expects that increasing the amount of water present will lead to less effective superexchange paths, and therefore lower transition temperatures.

On the other hand, this is not a monotonic trend, as can be seen by observing that the tetrahydrate has a lower T_c than the hexahydrate. It is likely that the reason for this lies less with the structural features of the two systems than with the fact that the zero-field splitting is substantially larger in the tetrahydrate.

Turning to the bromides, only the data as shown in Table 8.2 are available.

TABLE 8.2 Critical Temperatures of Hydrated Nickel Bromides.

	T_c,K	Reference
$NiBr_2 \cdot 6H_2O$	8.30	65
$NiBr_2 \cdot 2H_2O$	6.23	66
$NiBr_2$	60	67

Upon comparison with the analogous (and isostructural) chlorides, these compounds illustrate the useful rule of thumb that bromides <u>often</u> order at higher temperatures than chlorides (hexahydrate), as well as the violation of that rule (dihydrate). Two of the factors that influence the value of T_c are competing, for one expects the larger polarizability of bromide (over chloride) to lead to stronger superexchange interaction and thus a higher T_c, while the larger size of bromide causes the metal ions to be separated further which leads, in turn, to a lower T_c. A graphical correlation of these trends has been observed (67).

Aside from the anhydrous materials, there are not enough data available on the fluorides and iodides of nickel to discuss here.

It is convenient to separate the discussion of the dihydrates from that of the other hydrates, because the structural and magnetic behavior of the two sets of compounds are substantially different.

Nickel chloride hexahydrate is one of the classical antiferromagnets. Both the specific heat (61) and susceptibility (68) indicate a typical antiferromagnetic transition at 5.34 K. The crystal structure (69) shows

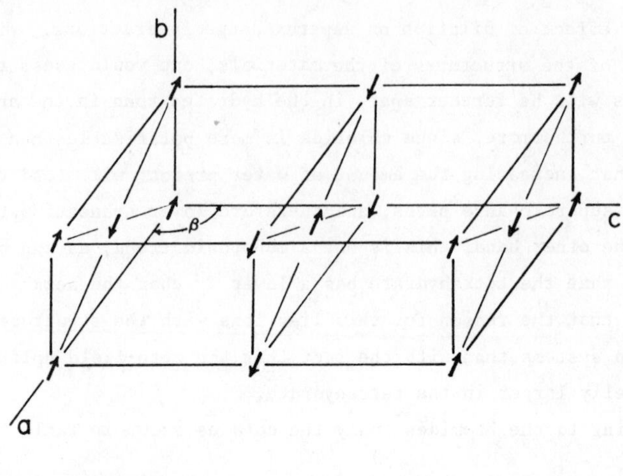

FIGURE 8.8 *Arrangement of magnetic moments in $NiCl_2 \cdot 6H_2O$ for the magnetic space group I_c2/c. The angle between the magnetic moment and the a axis is approximately 10°. From Ref. 72.*

that the monoclinic material consists of distorted trans- $[Ni(OH_2)_4Cl_2]$ units, hydrogen bonded together by means of an additional two molecules of water. The salt is isomorphous to $CoCl_2 \cdot 6H_2O$ yet the preferred axes of magnetic alignment in the two salts are not the same (68). There is relatively little magnetic anisotropy throughout the high-temperature (10-20 K) region (68,70), and, a point of some interest, neutron diffraction studies (71) show that the crystal structure is unchanged upon cooling the substance from room temperature to 4.2 K, which is below the critical temperature. As illustrated in Figure 8.8, additional neutron diffraction studies (72) have provided the magnetic structure, which consists of antiferromagnetic planes [001] with AF coupling between the planes. The result is that the magnetic unit cell is twice the size of the chemical cell, caused by a doubling along the c axis. The arrangement of the spins, or magnetic ordering, is the same as in $CoCl_2 \cdot 6H_2O$, except for direction. The reason for the differing orientation probably lies with the differing single-ion anisotropies, for the nickel system is an anisotropic Heisenberg system, while the cobalt compound, as discussed in Chapter VI-C, is an XY magnet. Early measurements of the susceptibilities disagreed somewhat with the neutron work on the location of the preferred axis, but recent remeasurements (70) show that the susceptibilities are consistent with the independent determination.

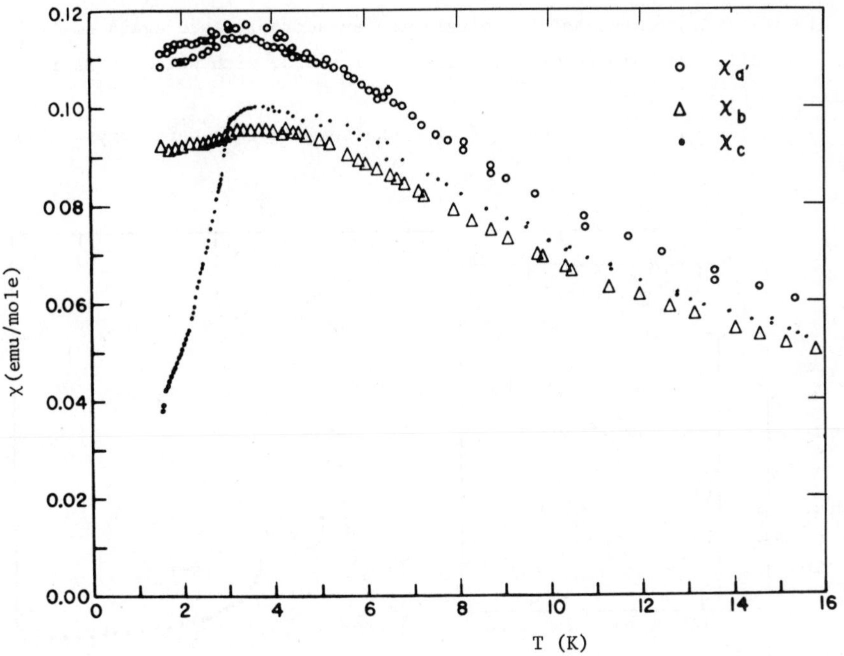

FIGURE 8.9 *The magnetic susceptibility of $NiCl_2 \cdot 4H_2O$ from 0 to 20 K along the* a'*,* b *, and* c *axes. From Ref. 62.*

What is interesting about $NiCl_2 \cdot 6H_2O$ from a chemist's viewpoint is that the anisotropies are low. The g value is normal, being isotropic at 2.22, but D/k = -1.5 ± 0.5 K and E/k = 0.26 ± 0.40 K; these results may be contrasted with those (62) for the very similar molecule, $NiCl_2 \cdot 4H_2O$. In this case, cis- $[Ni(OH_2)_4Cl_2]$ octahedra, which are quite distorted, are found. The compound is isomorphous with $MnCl_2 \cdot 4H_2O$, and in this case the easy axes of magnetization are the same. With the tetrahydrate, large paramagnetic anisotropy persists throughout the high-temperature region, as illustrated in Figure 8.9. With g again isotropic at 2.28, the zero-field splitting parameters in this case are D/k = -11.5 ± 0.1 K and E/k = 0.1 ± 0.1 K. This increased single-ion anisotropy is perhaps the governing factor in reducing T_c for $NiCl_2 \cdot 4H_2O$ below that of $NiCl_2 \cdot 6H_2O$; the antiferromagnetic exchange constants in the two crystals have been evaluated as almost equal. Portions of the magnetic phase diagrams of both salts are available (73).

The compound $NiBr_2 \cdot 6H_2O$ has also been investigated recently (65).

As was discussed above, a higher transition temperature than that of the chloride is observed, and the single-ion anisotropies are again small. All in all, the compound is much like $NiCl_2 \cdot 6H_2O$, with which it is isostructural.

Estimates of the internal fields can be made in several ways. The

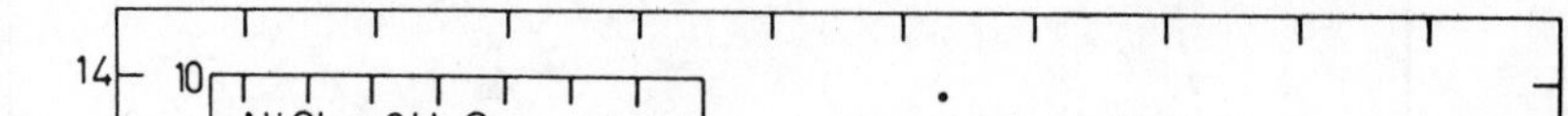

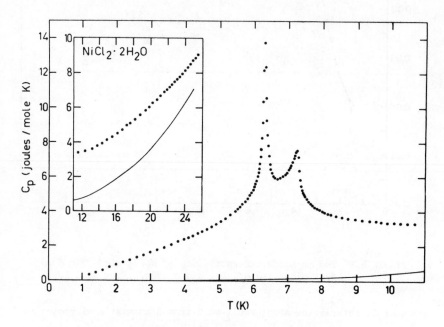

FIGURE 8.10 *Measured specific heat of $NiCl_2.2H_2O$. Note that the insert which shows the high-temperature results has an abscissa four times as coarse as the main graph. The solid line denotes the lattice estimate, $4.5 \times 10^{-4}\ T^3$ J/mole K. From Ref. 63.*

anisotropy constant K introduced in Chapt. VII-B, can be related to the zero-field splitting parameter D, by the relationship $K(T=0) = |D|NS(S-1/2)$ (70), where N is Avogadro's number and S is the spin of the ion. The anisotropy field can be defined as $H_A = K/M_S$, where $M_S = (1/2)Ng\mu_B S$. The exchange field is described by molacular field theory as $H_E = 2z|J|S/g\mu_B$. These fields have been estimated for both $NiCl_2 \cdot 6H_2O$ (70) and $NiBr_2 \cdot 6H_2O$ (65).

The compounds $NiCl_2 \cdot 2H_2O$ and $NiBr_2 \cdot 2H_2O$ differ appreciably from the other hydrates, primarily because they have the characteristic linear chain structure of trans- [$NiX_4(OH_2)_2$] units. The specific heats are of some interest because, as illustrated in Figure 8.10, the magnetic phase transition is accompanied by two sharp maxima for each compound. For the chloride, they occur at 6.309 and 7.258 K, while they occur at 5.79 and 6.23 K for the bromide. The implication is that there are two phase boundaries in the H-T plane, even at H=0. In fact, $NiCl_2 \cdot 2H_2O$ has a very complicated magnetic phase diagram, with five different regions (74). Neutron diffraction studies in zero-field as well as magnetization and susceptibility measurements show that at the lowest temperatures the system consists of ferromagnetic chains; neighboring chains in one crystallographic direction also have their moments pointing in the same direction, thus forming a ferromagnetic layer parallel to the ab plane. Successive layers in the c direction have opposite moments. As a result, although the predominant interactions are ferromagnetic in sign, the weak antiferromagnetic interaction between the planes leads to a gross antiferromagnetic configuration. The specific heat of $NiBr_2 \cdot 2H_2O$, which is assumed to be isostructural to the chloride, has likewise been interpreted (66) in terms of a large (negative) single ion anisotropy and ferromagnetic intrachain interaction. A theory has been presented (75) that allows spin reorientation as a function of temperature even in zero external field when large zero-field splittings are present which allow anisotropic exchange.

4. Hydrated Nickel Nitrates

A series of investigations (76-79) on the hydrates of nickel nitrate illustrates practically all of the problems in magnetism. That is, within the three compounds $Ni(NO_3)_2 \cdot 2H_2O$, $Ni(NO_3)_2 \cdot 4H_2O$, and $Ni(NO_3)_2 \cdot 6H_2O$, zero-field splittings, anisotropic susceptibilities, both short and long-range order, and metamagnetism are all to be found.

The different hydrates may be grown from aqueous solutions of nickel nitrate, with crystals of $Ni(NO_3)_2 \cdot 6H_2O$ appearing when the solution is kept at room temperature, the tetrahydrate at about $70^{\circ}C$, and the dihydrate at about $100^{\circ}C$. The hexahydrate undergoes 5 crystallographic phase transitions as it is cooled to low temperatures, so that its crystal structure in the helium region, where the magnetic measurements have been made, is unknown. The structure of the tetrahydrate is unknown, and the dihydrate forms layers such as is illustrated in Figure 8.11. Each nickel ion is bonded to two trans-water molecules, and to four nitrate oxygens that

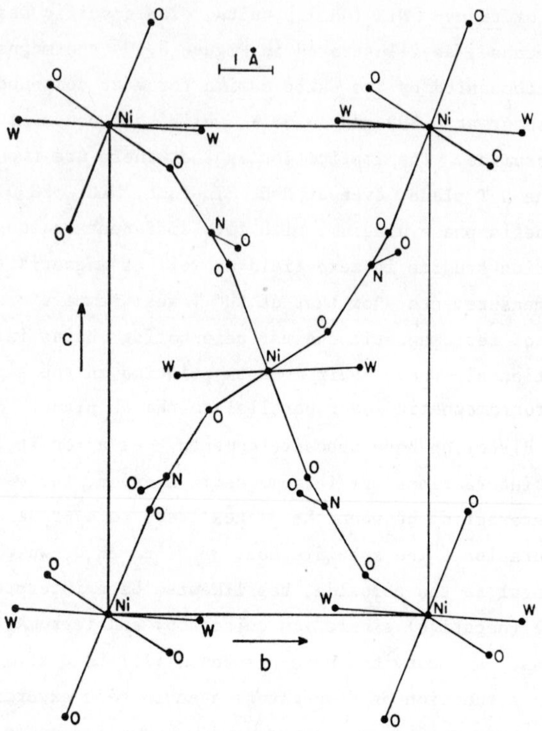

FIGURE 8.11 *Projection of the unit cell of nickel nitrate dihydrate on the bc plane. The nickel ions form a face-centered pattern. The symbol W is used to describe a water molecule. From Ref. 76.*

bridge to other nickel atoms.

Since these are nickel salts, of spin-1, zero-field splittings must be considered. Ignoring the rhombic (E) crystal field term for the moment, a negative axial (D) term puts a doubly-degenerate level below the singlet. A Schottky term is anticipated in the specific heat, but also, because the lowest level has spin-degeneracy irrespective of the size of D, magnetic ordering will occur at some temperature. This is in fact the situation with the dihydrate, where $D/k = -6.50$ K, and a λ-transition is found in the specific heat at 4.105 K.

But, in the hexahydrate (and also the tetrahydrate, which has been studied less extensively) the opposite situation prevails. That is, the zero-field splitting is positive and the singlet lies lowest. In the presence of exchange interactions that are weak compared to the zero-field splitting, all spin-degeneracy, and hence all entropy, are removed as the temperature is lowered towards 0 K, and the system cannot undergo long-range magnetic order (in the absence of a field). The situation is precisely the same as that discussed earlier for $Cu(NO_3)_2 \cdot 2\frac{1}{2}H_2O$ (Chapt. III). This argument has been put on a quantitative basis in the MF approximation

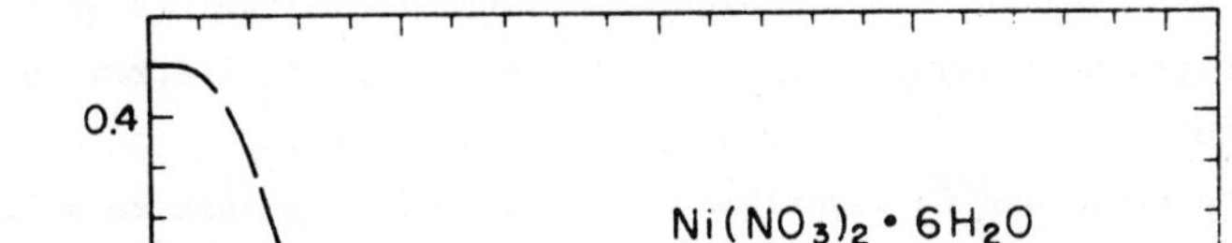

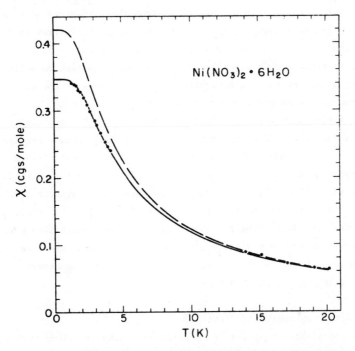

FIGURE 8.12 *Powder susceptibility of nickel nitrate hexahydrate as a function of temperature. Full line: theoretical curve, with g = 2.25, D/k = +6.43 K, E/k = +1.63 K, and A/k = +0.62 K. Dashed line: same, but with A = 0. From Ref. 79.*

by Moriya (44). Thus, the specific heat of $Ni(NO_3)_2 \cdot 6H_2O$ is described in terms of a Schottky function including both D and E. It was found that D/k = 6.43 K, and E/k = +1.63 K, which puts the three levels successively at 0, 4.80, and 8.06 K.

A magnetic transition is ruled out because none was observed above 0.5 K, all the entropy anticipated for a spin-1 ion was observed at higher temperatures from the Schottky curve, and thus exchange is quite small compared to the zero-field splittings. Magnetic interactions are not necessarily zero, on the other hand, although they are not evident in zero-field heat capacity measurements. Such subcritical interactions may be observable with susceptibility measurements, however, for they may contribute to the effective magnetic field at a nickel ion when an external field is applied, even if it is only the small measuring field in a zero-external field susceptibility measurement. This appears to be the case with $Ni(NO_3)_2 \cdot 6H_2O$, where the powder susceptibility is displayed in Figure 8.12, along with the susceptibility calculated from the parameters obtained from the heat capacity analysis. A good fit requires an antiferromagnetic molecular field constant, $A/k = 0.62$ K, where $A = -2zJ$.

The tetrahydrate is described by essentially the same zero-field splitting parameters as the hexahydrate. The g values of the three compounds are all normal, with a value of 2.25.

As mentioned above, the dihydrate orders at 4.1 K. A positive Curie-Weiss constant of 2.5 K suggests that the major interactions are ferromagnetic, and a MF model based on the structurally-observed layers has been proposed. If the layers are ferromagnetically coupled, with $2z_2J_2/k = +4.02$ K, and the interlayer interaction is weak and AF with $2z_1J_1/k = -0.61$ K, then a consistent fit of the susceptibility data is obtained. Furthermore, the large single-ion anisotropy, coupled with this magnetic anisotropy suggests that $Ni(NO_3)_2 \cdot 2H_2O$ should be a metamagnet, and this indeed proves to be the case. The phase diagram has been determined, and the tricritical point is at 3.85 K.

5. Tris Dithiocarbamates of Iron(III)

Six-coordinate complexes of iron(III) are usually either high spin (^6S ground state) or low spin ($^2T_2$). Several factors, such as the strength of the ligand field and the covalency, determine which configuration a particular compound will assume; moreover, whatever configuration that compound has, it is usually retained irrespective of all such external influences as temperature or minor modification of the ligand. The tris-(dithiocarbamates) of iron(III), $Fe(S_2CNR_2)_3$ are a most unusual set of compounds for, depending on the R group, several of them are high spin, several are low spin, and several seem to lie right at the cross-over point between the two configurations. While these are the best-known examples of this phenomenon, a review of the subject (80) describes several other such systems. Recent examples of other systems that can be treated formally as electronic isomers include molecules such as manganocene (81) and $Fe(S_2CNR_2)_2[S_2C_2(CN)_2]$ (82).

The dithiocarbamates and the measurements of their physical properties

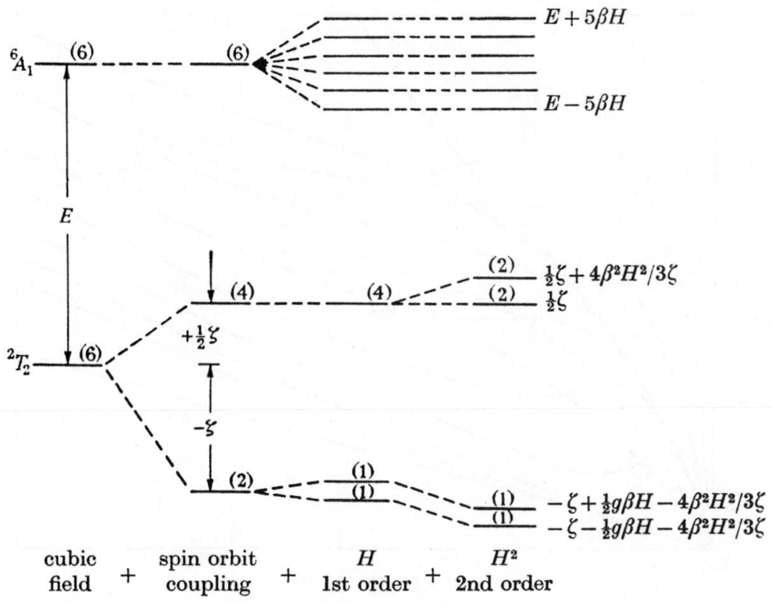

FIGURE 8.13 *Energy levels (not to scale) of configuration d^5 in the crossover region. From Ref. 80.*

have been reviewed at length elsewhere (80), so that only some of the magnetic results will be discussed here. The energy level diagram is sketched in Figure 8.13 for a situation in which the $^6A_1$ level lies E in energy above the $^2T_2$ state, where E is assumed, in at least certain cases, to be thermally accessible. The usual Zeeman splitting of the $^6A_1$ state is shown, but the splitting of the $^2T_2$ state is complicated by spin-orbit coupling effects, as indicated. The most interesting situations will occur when $E/kT \approx 1$, and the energy sublevels are intermingled.

The magnetic properties corresponding to this set of energy levels (80) are calculated from Van Vleck's equation as

$$\chi_M = N\mu_B{}^2\mu_{eff}{}^2/3kT \tag{8.5}$$

where

$$\mu_{eff}{}^2 = \frac{(3/4)g^2 + 8x^{-1}(1-e^{-3x/2}) + 105e^{-(1+E/\zeta)x}}{1 + 2e^{-3x/2} + 3e^{-(1+E/\zeta)x}} \tag{8.6}$$

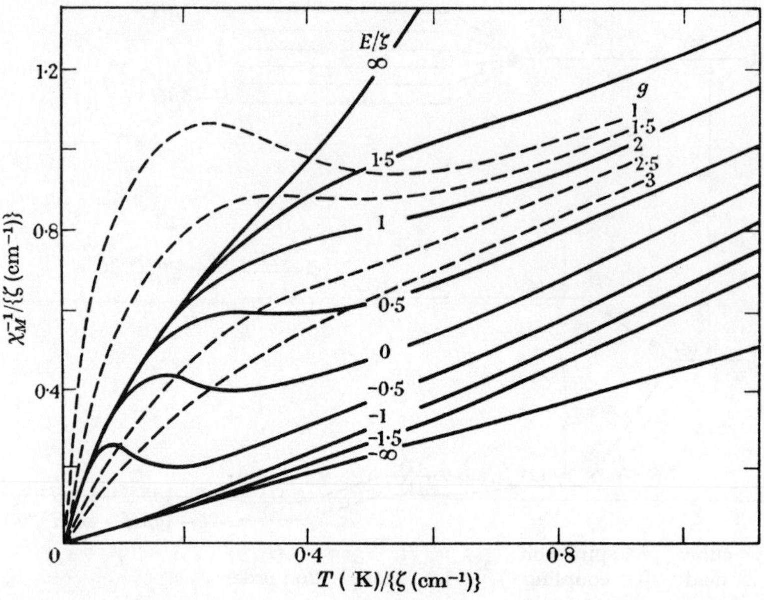

FIGURE 8.14 *Calculated values of* χ_M^{-1}. *Full lines:* $g = 2$ *with various values of* E/ζ. *Broken lines:* $E/\zeta = 1$ *with various values of* g. *From Ref. 80.*

and $x = \zeta/kT$, with ζ the one-electron spin-orbit coupling constant. Usual low spin behavior corresponds to large positive E, high spin, to large negative E, so that marked deviations from this behavior occur only when $|E/\zeta| \lesssim 1$. In practice, it is difficult to apply Eq. (8.5) over a wide enough temperature interval to evaluate all the parameters, especially since the spin-orbit coupling constant must often be taken as an empirical parameter; thermal decomposition and phase changes are among the other factors which limit the application of this equation.

The exceptional magnetic behavior of the crossover systems is best seen in the temperature dependence of the reciprocal of the molar susceptibility,

$$\chi_M^{-1} = 3kT/N\mu_B^2\mu_{eff}^2 \tag{8.7}$$

and several calculated values of χ_M^{-1} are represented (80) in Figure 8.14. The maxima and minima have obvious diagnostic value. A fuller discussion of the behavior of these curves has been given elsewhere (80). A selection of experimental data, along with fits to Eq. (8.5) (modified to account for different metal-ligand vibration frequencies in the two electronic states) is presented in Figure 8.15. It is remarkable how well a simple model accounts for such unusual magnetic behavior. Maxima and minima are

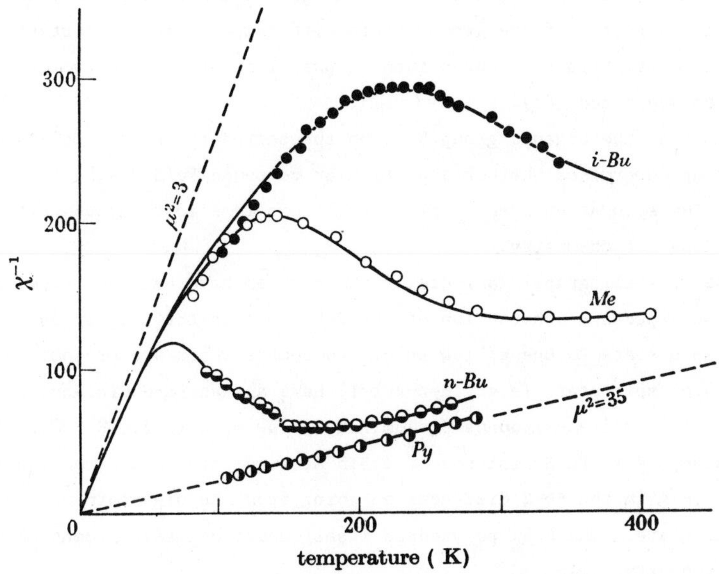

FIGURE 8.15 *Variation of χ^{-1} of $[Fe(S_2CNR_2)_3]$ with temperature. Me: R = methyl; n-Bu:R = n-butyl; i-Bu:R = i-butyl; Py:NR_2 = pyrrolidyl. From Ref. 80.*

observed (when the temperature range is appropriate), and the mean magnetic moment per iron atom rises with temperature from low spin toward high spin values. However, the relative positions of the $^6A_1$ and $^2T_2$ levels cannot be estimated beyond asserting that they must lie within about 500 cm^{-1} of each other. The marked discontinuity at 145 K in the susceptibility of the di-n-butyl derivative is due to a phase change of the solid.

It should be pointed out that similar data are obtained on these materials when they are dissolved in inert solvents, which shows that the reported effects are neither intermolecular in origin or due to some other solid state effect. Data on some 18 compounds in this series alone have been reported (80).

Although the magnetic aspects of dithiocarbamates appear to be understood on the whole, it should be pointed out that the more chemical aspects of this problem remain somewhat perplexing. For example, Mössbauer and proton NMR studies have failed to show evidence for the simultaneous population of two electronic levels (83), a result which would require a rapid crossover between the levels ($<10^{-7}$ sec). An alternative explanation, based on crystal field calculations, suggests (84) that the ground state may be a mixed-spin state, the variable magnetic moment being due to a change in character of the ground state with temperature. Important vibronic contributions to the nature of positions of the energy levels have also been suggested (85).

Although the organic group R is on the periphery of the molecule, this group determines whether a particular compound $Fe(S_2CNR_2)_3$ is high spin or low spin at a given temperature. This is a most unusual situation in coordination chemistry.

The crystallographic aspects of the problem have been reviewed (86, 87). One expects a contraction of the FeS_6 core in the transition from a high spin state to one of low spin. Compounds which are predominantly in the high spin state (e.g., R=n-butyl) have a mean Fe-S distance of about 2.41 Å, while compounds predominantly low spin (e.g., R = CH_3, R = C_6H_5) have a mean Fe-S distance of 2.315 Å, suggesting a contraction of about 0.08 Å in the Fe-S distances on going from the high-spin to the low-spin state. The FeS_6 polyhedron itself undergoes significant changes in its geometry, also.

6. Spin-3/2 Iron(III)

An interesting series of compounds is provided by the halo dithiocarbamates of iron, $Fe(X)[S_2CNR_2]_2$, for they exhibit the unusual spin state of 3/2 for iron(III). In octahedral stereochemistry, as was illustrated above, iron(III) must always have spin of 5/2 (so-called spin-free), or 1/2 (spin-paired), each with its own well-defined magnetic behavior. Since these spin states also apply to tetrahedral stereochemistry, although no spin-1/2 tetrahedral compounds of iron(III) have yet been reported, it is impossible to have a ground state with S = 3/2 in cubic coordination. But, in lower symmetry environments, such as in a distorted five-coordinate structure, the S = 3/2 state can in fact become the ground state, and that happens with these compounds.

The compounds are easily prepared (88) from the corresponding tris-(dithiocarbamates) by the addition of an aqueous hydrohalic acid to a solution of $Fe[S_2CNR_2]_3$ in benzene. The compound where R = C_2H_5 and X = Cl has been studied the most widely, but a variety of compounds have

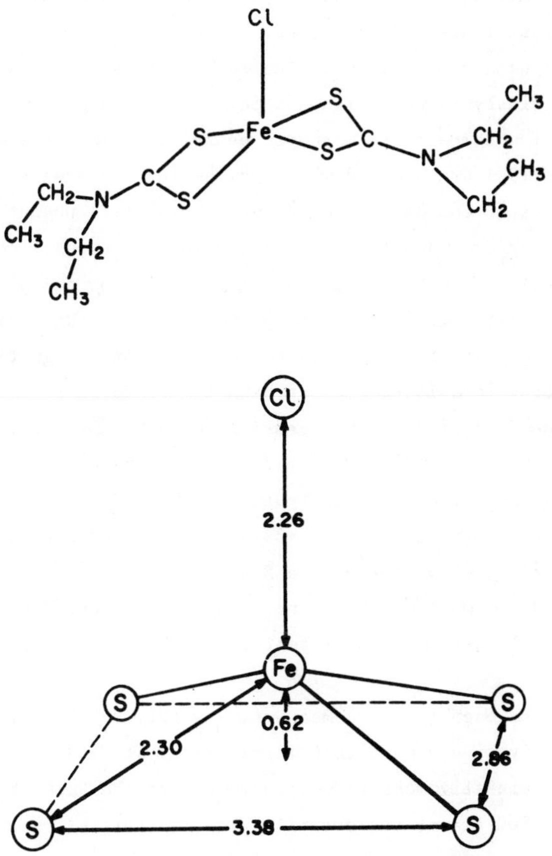

FIGURE 8.16 *Molecular geometry of Fe(dtc)$_2$X. From Ref. 94.*

been prepared and studied with, for example, X = Cl, Br, I or SCN, and R = CH_3, C_2H_5, i-C_3H_7, etc. All are monomeric and soluble in organic solvents, and retain a monomeric structure in the crystal (88-91). The structure of a typical member of the series is illustrated in Figure 8.16, where the distorted tetragonal pyramidal coordination that commonly occurs may be seen. Thus, the four sulfur atoms form a plane about the iron atom which lies, however, 0.63 Å above the basal plane. The Fe-X distances are 2.26 Å (Cl), 2.42 Å (Br) and 2.59 Å (I). In solution at ambient temperatures the compounds exhibit a magnetic behavior typical of a S = 3/2 system (88). All available data are consistent with the fact that the orbitally non-degenerate $^4A_2$ state is the ground state.

As usual, the $^4A_2$ state is split by the combined action of spin-orbit coupling and crystal field distortions. Ignoring the small rhombic (E) terms that have been reported (92), the zero-field splittings 2D for these compounds vary widely in both sign and magnitude, and of course this determines the low-temperature magnetic behavior. Since the g-values of these compounds (93) are about 2 when considering the true spin of 3/2, the situation is similar to that of tetrahedral cobalt(II) when the zero-field splittings are large. Thus, $|\pm 1/2>$ is low when D is positive, and $g'_{\parallel} = 2$, $g'_{\perp} = 4$; conversely, when D is negative, $|\pm 3/2>$ is low and $g'_{\parallel} = 6$, $g'_{\perp} = 0$. The zero-field splittings in these compounds have been determined from the measurement of the far IR spectra in the presence of a magnetic field (92) and from paramagnetic anisotropies at high temperatures (94). Choosing the Fe-X bond as the direction of the quantization axis (94) in the ethyl derivatives, $D = +1.93 \pm 0.01$ cm^{-1} for the chloro compound, and $D = +7.50 \pm 0.10$ cm^{-1} for the bromo derivative. Large E (rhombic) terms have also been reported; writing $\delta = 2(D^2+3E^2)^{1/2}$, reported zero field splittings δ for the ethyl series are 3.4 cm^{-1} (Cl), 16 cm^{-1} (Br), and 19.5 cm^{-1} (I) (94). On the other hand, the magnetic and Mössbauer results on $Fe(Cl)(S_2CNEt_2)_2$ are more consistent if the magnetic z axis lies in the FeS_4 plane (95).

Mössbauer and magnetization measurements (91, 96-99) show that the chloro ethyl derivative orders ferromagnetically at 2.43 K and that the intramolecular spin alignment is perpendicular to the Fe-Cl bond and parallel to the FeS plane; the overall spin-structure of the unit cell is not yet available, however. While the iodo ethyl compound with $D > 0$ appears to be antiferromagnetic with $T_c = 1.95$ K, it is interesting to note that the bromo ethyl derivative remains paramagnetic down to 0.34 K. The bromo(morphylyldithiocarbamate) analog orders at 3.50 K and should be an XY magnet, but the compound exhibits polymorphism, and variable solvent adducts form; iodo(pyrrolidinedithiocarbamato)iron(III) orders at 2.16 K (100). No specific heat or single crystal susceptibility data are available yet, but this series of compounds should be interesting to examine further at low temperatures.

7. Manganous Acetate Tetrahydrate

The structure (101,102) of $Mn(CH_3COO)_2 \cdot 4H_2O$ is illustrated in Figure 8.17, where it will be observed that the crystal consists of planes containing trimeric units of manganese atoms. The manganese atoms are

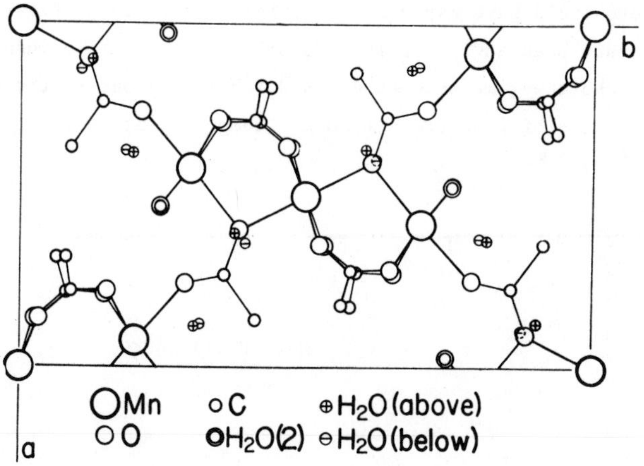

FIGURE 8.17 *The structure of $Mn(CH_3COO)_2 \cdot 4H_2O$. The projection is along the perpendicular c* direction. Two water molecules in the Mn_2 coordination octahedron which are shown superimposed are actually one above and one below the Mn plane. From Ref. 102.*

inequivalent, for the central one lies on an inversion center, and is bridged to both of its nearest neighbors by two acetate groups in the same fashion as in copper acetate. However, in addition to the water molecules in the coordination spheres, there is also another acetate group present which bonds entirely differently, for one of the oxygen atoms bridges Mn_1 and Mn_2, while the second oxygen atom of the acetate group forms a longer bridging bond to an Mn_2 of an adjoining trimer unit. The net result of these structural interactions is that there is a strong AF interaction within the trimer, and a weaker interaction between the trimers. There appears to be considerable short range order in the compound above the long-range ordering temperature of 3.18 K.

The inverse powder susceptibility (103) shows substantial curvature throughout the temperature region below 20 K. Between 14 and 20 K, apparent Curie-Weiss behavior is observed, $\chi = 3.19/(T+5.2)$ emu/mole, but the Curie constant is well below the value 4.375 emu K/mole normally anticipated for $S = 5/2$ manganese. The large Curie-Weiss constant, and the curvature in χ^{-1} below 10 K, coupled with the small Curie constant, suggest that there is probably also curvature in the hydrogen region, and that in fact true Curie-Weiss behavior will be observed only at much higher temperatures. This is one piece of evidence which suggests the presence of substantial short-range order.

The zero-field heat capacity (104) is illustrated in Figure 8.18, where the sharp peak at $T_c = 3.18$ K is only one of the prominent features. The broad, Schottky-like peak at about 0.7 K may be due to the presence of inequivalent sets of ions present in the compound; it could arise if one

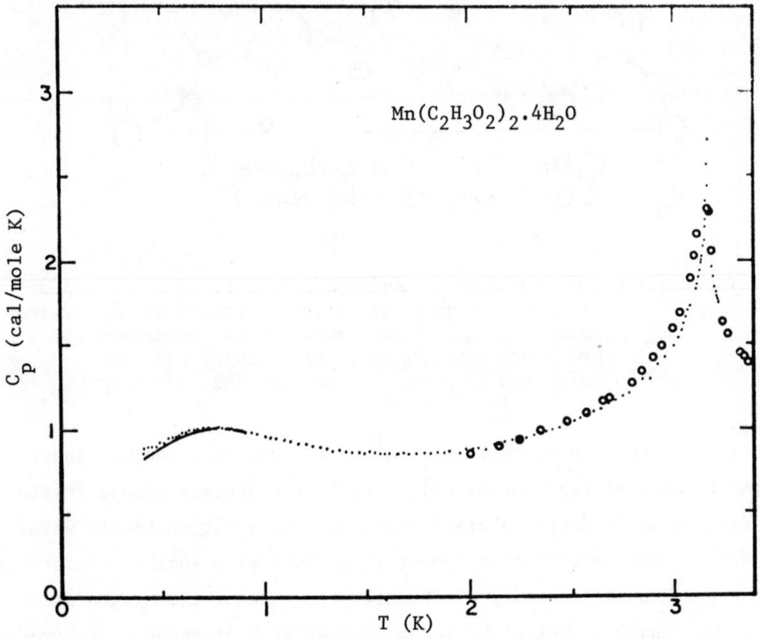

FIGURE 8.18 *Heat capacity C_p of powdered $Mn(CH_3COO)_2 \cdot 4H_2O$ in "zero" applied field below 3.4 K.*

sublattice remains paramagnetic while the other sublattice(s) become ordered, but a quantitative fit of this portion of the heat capacity curve is not yet available. The other, unusual feature of this heat capacity curve is that the specific heat is virtually linear between 5 and 16 K. Thus, the heat capacity cannot be governed by the familiar relation, $C = aT^3 + b/T^2$, which implies that either the lattice is not varying as T^3 in this temperature region, or that the magnetic contribution is not varying as T^{-2}, or both. From the fact that less than 80 % of the anticipated magnetic entropy is gained below 4 K, at the least, these facts suggest the presence of substantial short-range order.

Unusual anisotropy was observed (103) in the single crystal suscepti-

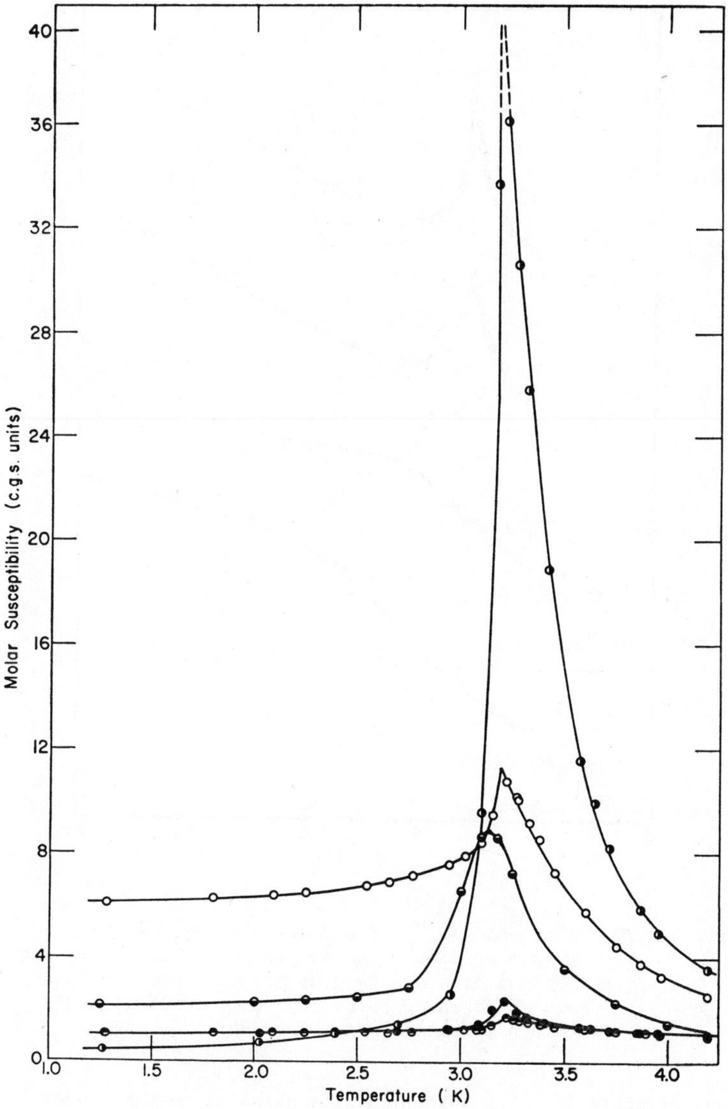

FIGURE 8.19 *Powder and single crystal susceptibilities of $Mn(CH_3COO)_2.4H_2O$. o, b axis; ● and •, c^* axis; ◐, a axis; ◒, powder. From Ref. 103.*

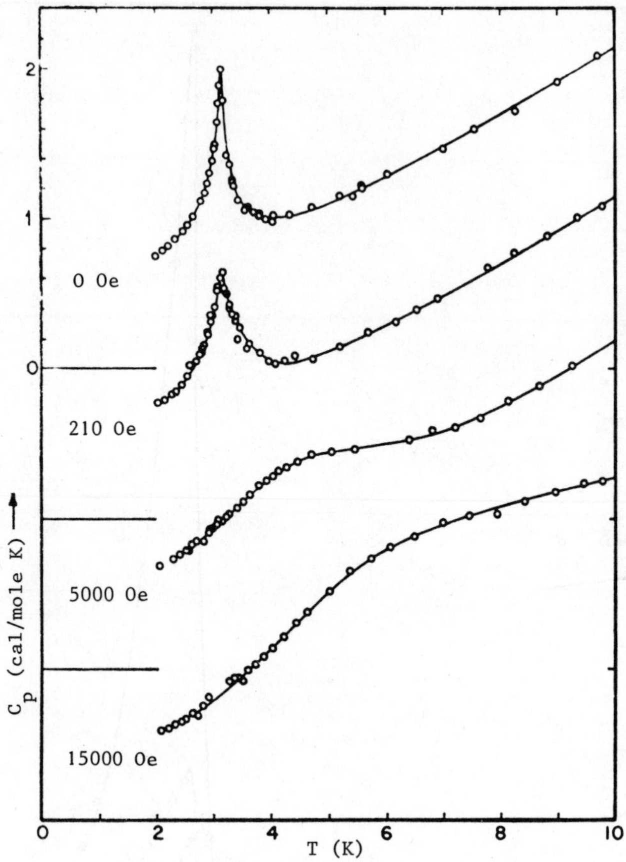

FIGURE 8.20 *Heat capacity C_p of powdered $Mn(CH_3COO)_2 \cdot 4H_2O$ in several applied magnetic fields. The zero of the vertical axis has been displaced downwards to separate different curves. The same vertical scale applies to each curve. From Ref. 104.*

bilities, Figure 8.19. The susceptibility parallel to the monoclinic a-axis rises sharply to an unusually large value while in the c* direction (a, b, c* are a set of orthogonal axes used in the magnetic studies in the $P2_1/a$ setting; the crystallographers (101,102) prefer the $P2_1/c$ setting) a similar but smaller peak is observed. The third, orthogonal susceptibility is essentially temperature-independent in this region, the bump in χ_b probably being due to misalignment. While it is clear from these data that a magnetic phase transition occurs in $Mn(OAc)_2 \cdot 4H_2O$ at 3.18 K, the

behavior of χ_a alone suggests that the phase transition is not the usual paramagnetic to antiferromagnetic one.

The compound is unusually sensitive to external magnetic fields, Figure 8.20 . The λ-peak in the specific heat broadens and shifts to

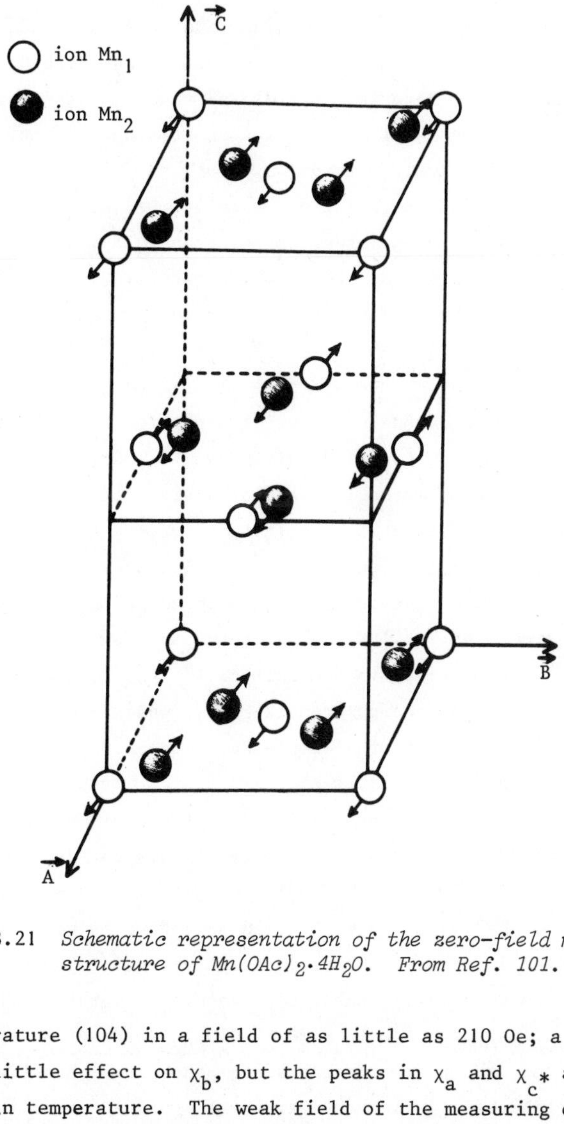

FIGURE 8.21 *Schematic representation of the zero-field magnetic structure of* $Mn(OAc)_2 \cdot 4H_2O$. *From Ref. 101.*

higher temperature (104) in a field of as little as 210 Oe; a field of 1000 Oe has little effect on χ_b, but the peaks in χ_a and χ_{c^*} are reduced and shifted in temperature. The weak field of the measuring coils was found to influence χ_a, and indeed a magnetic phase transition is caused by an external field of a mere 6 Oe (105,106).

The saturation magnetization of the compound (106) is only one-third as large as anticipated for a normal manganese salt. This would follow if the exchange within the Mn_2-O-Mn_1-O-Mn_2 groups is antiferromagnetic and relatively large compared to the coupling between such groups in the same plane or between planes. The neutron diffraction work (101) shows that the interactions within the Mn_2-O-Mn_1-O-Mn_2 units is AF, but ferromagnetic between different groups in the same plane and AF between adjacent planes. The zero-field magnetic structure deduced from the neutron work is illustrated in Figure 8.21. Both the crystal and magnetic structures are consistent with the existence of substantial short range magnetic order.

NMR studies (105) are consistent with all these data. The transition at 6 Oe with the external field parallel to the a-axis is like a metamagnetic one, short range order persists above T_c, and the saturated paramagnetic state is found at about 140 Oe (at 1.1 K). No exchange constants have as yet been extracted from the data.

8. $[M(C_5H_5NO)_6](ClO_4)_2$

This series of compounds with pyridine N-oxide ligands is of interest for several reasons. The compounds with M=Mn, Fe, Co, Ni, Cu, and Zn are known to be isomorphous, which is especially significant because the zinc compound provides a diamagnetic host lattice for EPR and other studies, and the copper compound (at least at room temperature) presumably then does not display the distorted geometry that is common with so many other compounds. The structure (107,108), illustrated in Figure 8.22, is rhombohedral with but one molecule in the unit cell. The result is that each metal ion has six nearest-neighbor metal ions, which in turn causes the lattice to approximate that of a simple cubic one (109).

The metal ions in this lattice attain strict octahedral symmetry, yet they display relative large zero-field splittings. Thus, the O-Ni-O angles are $90.3(1)^o$ and $89.7(1)^o$ in $Ni(C_5H_5NO)_6(BF_4)_2$ and the O-Co-O angles are $89.97(4)^o$ and $90.03(4)^o$ in $[Co(C_5H_5NO)_6](ClO_4)_2$. Yet, in the nickel perchlorate compound (41), the zero-field splitting is very large, 6.26 K, and the parameter δ (Section B8) is relatively large (-500 to -600 cm^{-1}) in the cobalt compound (110); the latter result is obtained from the highly anisotropic g-values of $[Co,Zn(C_5H_5NO)_6](ClO_4)_2$, with $g_{\parallel} = 2.26$, $g_{\perp} = 4.77$. Similarly, in the isomorphous manganese compound, the zero-field splitting parameter, D, takes (111) the very large value of 410 gauss (0.055 K).

The antiferromagnetic ordering behavior observed with the $[M(C_5H_5NO)_6](ClO_4)_2$ molecules is quite fascinating, especially as it occurs

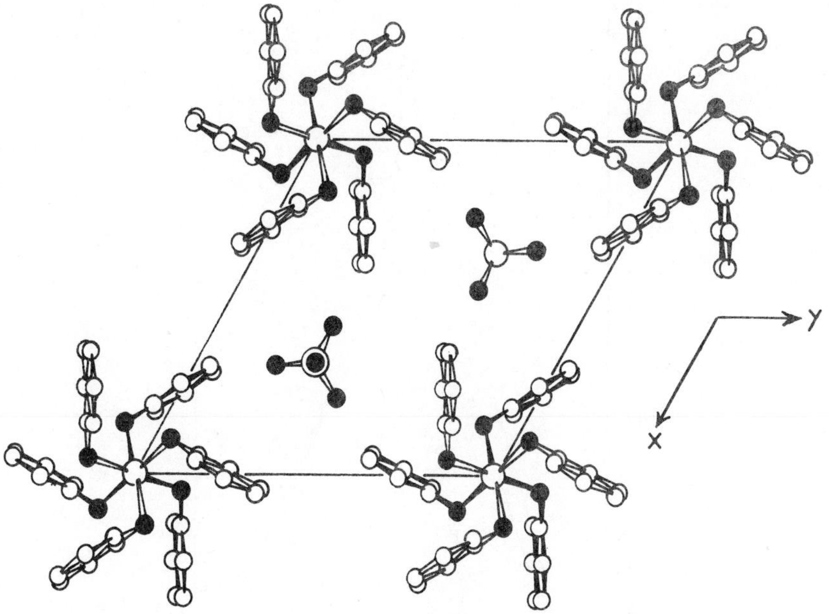

FIGURE 8.22 *A projection along the c-axis of one layer of the hexagonal unit cell of* $[Co(C_5H_5NO)_6](ClO_4)_2$. *The cobalt ions are on the corners (large open circles) and are octahedrally surrounded by the oxygens (filled circles) belonging to the* C_5H_5NO *groups. It should be noted that the cobalt ions shown are next-nearest magnetic neighbors to one another. From Ref. 103.*

at what is, at first glance, relatively high temperatures. Thus, the cobalt molecule orders at 0.428 K (109). When ClO_4^- is replaced by BF_4^- in these substances, the crystal structures remain isomorphous (107), and the magnetic behavior is consistent with this. Thus, $[Co(C_5H_5NO)_6](BF_4)_2$ orders at 0.357 K, and on a universal plot of C/R vs. $kT/|J|$, the data for both the perchlorate and fluoborate salts fall on a coincident curve (109), as illustrated in Figure 8.23. The nitrate behaves similarly. These molecules are the first examples of the simple cubic, S=1/2, XY magnetic model.

The situation changes in a remarkable fashion with the copper analogs (112). Though they are isostructural with the entire series of molecules at room temperature, the copper members apparently distort as they are cooled. This is consistent with the usual coordination geometries found with copper, but what is especially fascinating is that the perchlorate

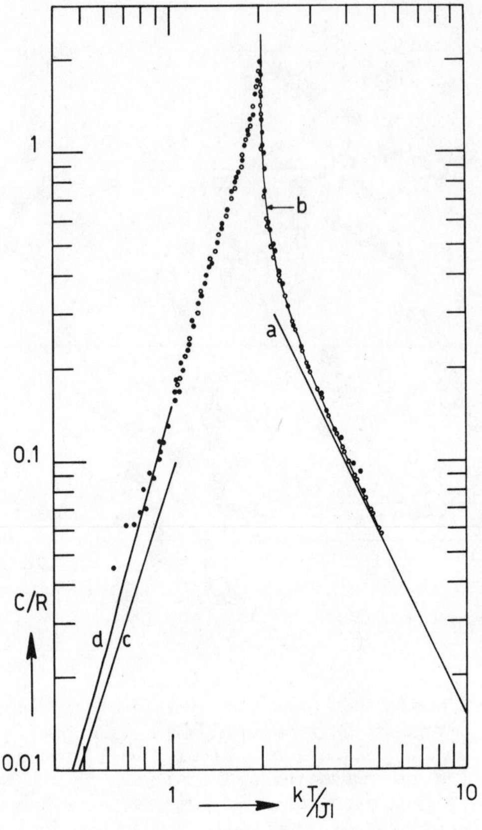

FIGURE 8.23 *Specific heat data for* $[Co(C_5H_5NO)_6](BF_4)_2$ *(open circles) and* $[Co(C_5H_5NO)_6](ClO_4)_2$ *(filled circles) plotted vs. relative temperature,* $kT/|J|$. *Curves a-d are theoretical predictions for the s.c. XY model with* $S = 1/2$. *From Ref. 109.*

becomes a one-dimensional magnetic system at helium temperatures, while the fluoborate, which must distort differently, behaves as a two-dimensional magnet.

In the case of $[Fe(C_5H_5NO)_6](ClO_4)_2$, yet another interesting phenomenon is observed. A zero-field splitting large even for this ion is implied by the Mössbauer measurements (31), for the lowest excited component of the ^5D state is estimated to be about 154 K above the ground state. This leaves an effective doublet ground state, with $g_{\parallel} = 9.0$ and $g_{\perp} = 0.6$, suggesting that this molecule ought to be a 3-d, $S' = 1/2$ Ising magnet.

In fact, this is substantiated by heat capacity measurements, with a very sharp λ-transition being found at about 0.7 K (113).

The compound $[Mn(C_5H_5NO)_6](BF_4)_2$ orders at 0.16 K (114). Though this is not an extraordinary low temperature, this means for this particular compound that magnetic exchange interactions are comparable in magnitude with the zero-field splittings, dipole-dipole interactions, and even nuclear hyperfine interactions. The net result is that an odd shaped peak is observed and a separation of the several contributions can only be made with difficulty.

The large zero-field splitting (6.26 K) observed with $[Ni(C_5H_5NO)_6]$-$(ClO_4)_2$ has already been discussed (Section B9 and References 27, 41). The antiferromagnetic interactions are subcritical ($zJ/k = -1.5 \pm 0.5$ K) so that magnetic ordering cannot occur at any temperature at zero external magnetic field. But, since this substance belongs to a rhombohedral (Z = 1)or hexagonal (Z = 3) crystal system, it presents a situation similar to that described earlier (Chapt. V-G) for $Cu(NO_3)_2 \cdot 2\frac{1}{2}H_2O$, in that an external field can be applied parallel simultaneously to the three-fold axes of all the molecules in the sample. The ground state then becomes degenerate as the lower two levels cross (Fig. 3.11) at the field $H_{eff} = (D + zJ)/g\mu_B$, and then spin ordering becomes possible. Theoretical work (44,115) shows that the interactions between the effective spins $S' = 1/2$ exhibit XY anisotropy ($J_z = 1/2\ J_\perp$), and a long-range ordered state is present below $T_{Cmax} = z|J|\cdot/k$, only between the critical fields $H_{c1} = (D-2z|J|)/g\mu_B$ and $H_{c2} = (D+4z|J|)/g\mu_B$, at $T = 0$ K. Furthermore, this ordering is characterized by a spontaneous magnetization or alignment of spins perpendicular to the external field.

Such a situation has indeed been realized with $[Ni(C_5H_5NO)_6](ClO_4)_2$ (116). The susceptibilities parallel to the 3-fold axis in fields between 25 and 70 kOe, and at constant temperatures between 0.08 and 1 K remain more-or-less constant; the molecular field theory predicts that the susceptibility at $T = 0$ K will be constant for fields between the two critical fields. The lowest temperature curve resembles this behavior quite well, which reflects the perfect 3-d character of the exchange interactions in this salt. Furthermore, $\chi(T)$ in constant fields between 29 and 62 kOe resembles the zero-field perpendicular susceptibility of a normal, 3-d antiferromagnet, as would be expected since the external field is perpendicular to the easy plane of the effective spins. The maximum critical temperature observed is 0.80 K, with the external field at about 49 kOe; the deviation of T_c from the value above 1 K predicted by the molecular field theory is not unexpected. These results are confirmed by specific heat measurements on an oriented sample in a field (117).

The resulting phase diagram is illustrated in Figure 8.24. The system is antiferromagnetic within the semi-circular region, and paramagnetic outside.

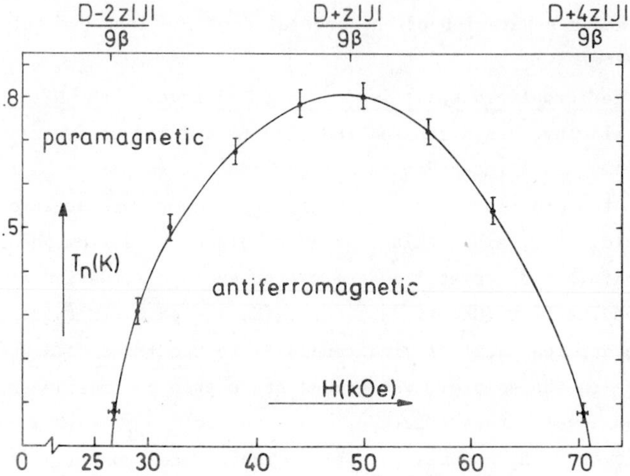

FIGURE 8.24 *Phase diagram of the field-induced antiferromagnetically ordered state of* $[Ni(C_5H_5NO)_6](ClO_4)_2$. *From Ref. 116.*

In order to determine the superexchange path in this series of molecules, it is necessary to examine the crystal structure with care. This is true, of course, in any system but especially so here, where a naive approach would suppose that the pyridine rings would effectively insulate the metal ions from one another and cause quite low ordering temperatures. Figure 8.25 illustrates several facets of the likely superexchange paths in these molecules. The rhombohedral unit cell formed by the metal ions is illustrated and closely approximates a simple cubic lattice. A reference metal ion is connected to its six nearest magnetic neighbors at a distance of about 9.6 Å by equivalent superexchange paths, consisting of a nearly collinear Co-O-O-Co bond, in which the Co-O and O-O distances are about 2.1 Å and 5.6 Å, respectively. It is important to note that the pyridine rings are diverted away from this bond so that one expects the superexchange to result from a direct overlap of the oxygen wave functions.

A nice illustration of the importance of the superexchange path, or rather, the lack of a suitable superexchange path, is provided by a recent study (118) of the hexakis-imidazole- and antipyrine cobalt(II) complexes. The shortest direct metal-metal distances are long, being, for example, 8.68 Å in $[Co(Iz)_6](NO_3)_2$; furthermore, while effective paths such as a Co-O-O-Co link can be observed in the antipyrine compound, the angles are sharper than in the pyridine N-oxide compound, and the O-O distance is also longer (7.6 Å vs. 5.6 Å for the C_5H_5NO salt). Neither compound orders magnetically above 30 mK.

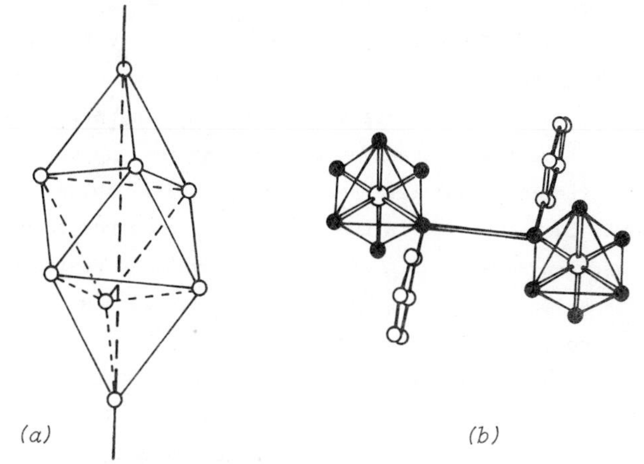

FIGURE 8.25 (a) *The rhombohedral cell formed by the cobalt ions in* $[Co(C_5H_5NO)_6](ClO_4)_2$.
(b) *The superexchange path connecting cobalt ions that are nearest magnetic neighbors.*
From Ref. 109.

The molecule $[Co(DMSO)_6](ClO_4)_2$, where DMSO is $(CH_3)_2SO$, behaves similarly magnetically (119); although the structure is as yet unknown, the coordination sphere is expected to be similar to that found with the pyridine N-oxides. However, it does not order above 35 mK.

9. NiX_2L_2

Many of the principles discussed earlier in this book come together in the series of compounds of stoichiometry MX_2L_2, where M may be Cu, Ni, Mn, Co, or Fe; X is Cl or Br; and L is pyridine (py) or pyrazole (pz). All the molecules appear to be linear chains in both structure and magnetic behavior, with <u>trans</u>-MX_4L_2 coordination spheres. The dihalo-bridged chains are formed by edge-sharing of octahedra; the structure of this isostructural series of molecules was illustrated for the Ising chain α-$CoCl_2 \cdot 2py$ in Figure 6.1. It was pointed out in Chapt. VI-B that $CoCl_2 \cdot 2L$, where L is H_2O or pyridine, exhibits a large amount of short-range order, but that the effect was enhanced when the small water molecule was replaced by the larger pyridine molecule. Similarly, large amounts of short-range order as well as unusually large zero field splittings are also exhibited by the other members of this series. The discussion will be limited to the nickel compounds.

Susceptibilities of the compounds NiX_2L_2, X = Cl, Br; L = py, pz, have been reported under several conditions (120). Unfortunately, this series of molecules does not form single crystals with any degree of ease, and only powder measurements are available. The magnetic behavior is unusual enough to be easily observed in this fashion, however, and the results are confirmed by specific heat studies (121), so that a great deal of confidence can be placed in the work. The compounds obey the Curie-Weiss law between 30 and 120 K with relatively large positive values of θ of 7 to 18 K, depending on X and L. The Curie constants are normal for nickel(II), and since it was known that a chain structure obtains, the θ parameters were associated with the intrachain exchange constant, J. Deviations from the Curie-Weiss law were observed below 30 K, which are due to the influence of both intrachain exchange and single-ion anisotropy due to zero-field splittings.

The analysis of the susceptibility data is hindered by the lack of suitable theoretical work for chains of spin-1 ions. Writing the Hamiltonian as

$$\mathcal{H} = g\mu_B \vec{H}\cdot\vec{S} + D[S_z^2 - (1/3)S(S+1)] + A\,\vec{S}\cdot\langle\vec{S}\rangle, \qquad (8.8)$$

a molecular field term, with A = 2zJ, is included to account for the intrachain interaction. The above Hamiltonian may be solved approximately for T larger than either of the parameters D/k or A/k, and, after averaging the three orthogonal susceptibilities in order to calculate the behavior of a powder, one finds (120)

$$\theta = 4zJ/3k$$

and so ferromagnetic intrachain constants of 2 to 7 K were obtained, in addition to very large zero-field splittings of the order of −25 K.

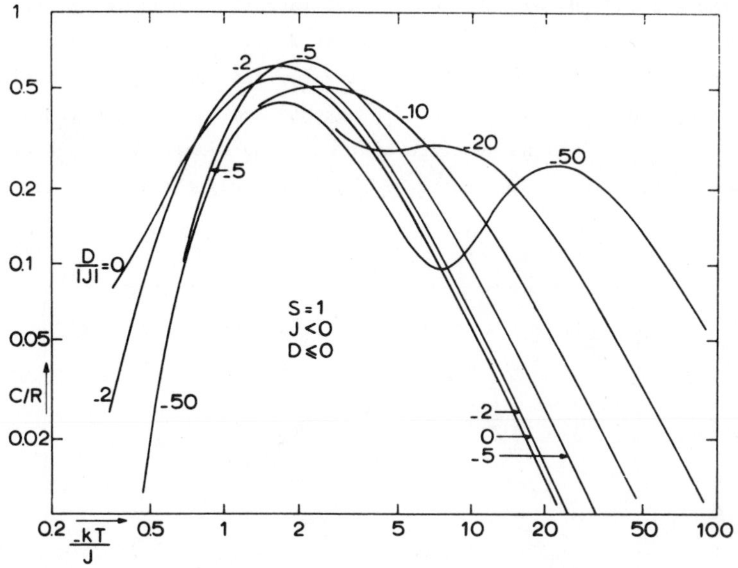

FIGURE 8.26 *Heat capacities of antiferromagnetic S = 1 chains with isotropic interaction and negative D terms. For large D, the extrapolated results approach the sum of a Schottky anomaly (due to the D term) and an Ising anomaly for magnetic interaction in the lower doublet. From Ref. 122.*

At low temperatures, and at low fields, the powder susceptibility exhibits a maximum, and then approaches zero value at 0 K. The temperatures of maximum χ, T_m, are about 3 to 7 K, and were assigned as critical temperatures; comparison with specific heat results suggests that T_c is actually slightly below T_m, as discussed in Chapt. V-D. A weak antiferromagnetic interchain interaction that leads to an antiferromagnetically ordered state would cause this behavior.

All these results are confirmed by the specific heat data (121). The broad peaks in the magnetic heat capacity which are characteristic of one-dimensional ordering were observed, and transitions to long-range order, characterized by λ-like peaks, were observed; a double peak was observed for $NiCl_2 \cdot 2py$. The ordering temperatures are 6.05 K ($NiCl_2 \cdot 2pz$); 3.35 K ($NiBr_2 \cdot 2pz$); 6.41 and 6.750 K ($NiCl_2 \cdot 2py$); and 2.85 K ($NiBr_2 \cdot 2py$).

There are several features of these investigations that make these compounds of more than passing interest. The first is that since the compounds contain nickel, which is a spin-1 ion, zero-field splittings would be expected to, and do, complicate the specific heat behavior. In fact, these compounds exhibit some of the largest zero-field splittings yet observed being, for example, -27 K for $NiCl_2 \cdot 2py$, and -33 K for $NiBr_2 \cdot 2pz$. The parameter D/k is negative for the four compounds which places a spin-doublet as the lowest or ground state. The problem then arises of calculating the specific heat of a one-dimensional magnetic system as a function of the ratio D/J, and this was the inspiration for the

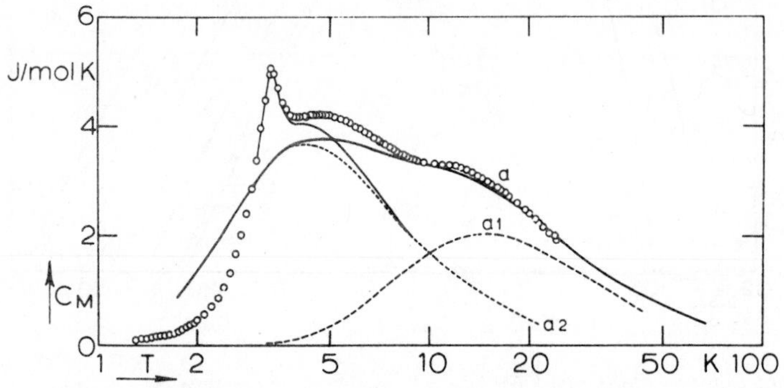

FIGURE 8.27 *The magnetic specific heat of $NiBr_2 \cdot 2pz$ as a function of temperature: 0, experimental results; ---a1, Schottky curve for independent Ni(II) ions with single-ion anisotropy parameter D/k = -31 K. ----a2, Ising linear-chain model S = 1/2; J/k(S = 1/2) = 10.5 K; ——a, the sum curve of a1 and a2. This curve coincides for the greater part with the curve for the Heisenberg linear-chain model with uniaxial single-ion anisotropy as calculated for spin 1 by Blöte; J/k = 2.7 K, D/k = -33 K. From Ref. 121.*

work of Blöte (114). A typical set of his results is illustrated in Figure 8.26, where it will be observed that as $D/|J|$ becomes large and negative, that two broad peaks are to be found in the specific heat. One is due to the magnetic chain behavior, and the second is due to the Schottky term. That these contributions are in fact additive when they are well-separated on the temperature axis (i.e., when $D/|J|$ is large) was illustrated by the data on $NiBr_2 \cdot 2pz$, Figure 8.27. An equally good fit to the data was obtained by either fitting the results to the complete curve of Blöte for the Heisenberg linear chain model with uniaxial single-ion anisotropy, or by simply summing the linear chain and Schottky contributions. It should be pointed out that when $|D/J|$ is as large as it is, about 12, in this compound, that the exchange interaction occurs between ions with effective spin-doublet ground states. The situation was described in Chapt. III-F and corresponds to an Ising $S' = 1/2$ system. Care must be used in comparing the magnetic parameters obtained by the analyses from the different points of view, since the J/k obtained from the $S' = 1/2$ formalism will be four times larger than that corresponding to the $S = 1$ Hamiltonian.

The fact that the $S' = 1/2$ Ising ion has g values of $g_{\|} \approx 4.4$ and $g_{\perp} \approx 0$ gives rise to the other interesting feature of $NiBr_2 \cdot 2pz$, and that is that this Ising nature, along with the ferromagnetic intrachain interaction, causes it to be a metamagnet. The specific heat of a powdered sample was measured in a field of 5 kOe; since $g_{\perp} \approx 0$, the (unavailable) single crystal measurements were not needed. The λ-like anomaly disappeared, the maximum value of the specific heat increased and shifted to a higher temperature. Furthermore the susceptibility as a function of field (120) has a maximum at the critical field for the AF → Metamagnetic transition, and then maintains a constant value. Not only do these results confirm the metamagnetic behavior, but also they confirm the unusually large zero-field splittings.

10. $[(CH_3)_3NH_3]MX_3 \cdot 2H_2O$

This is an extensive series of interesting magnets (115). For the most part, they show a large degree of linear chain behavior, but a number of other features are of interest. In particular, they exhibit remarkably well a correlation of magnetic properties with structure.

The crystal structure of $[(CH_3)_3NH]MnCl_3 \cdot 2H_2O$ is illustrated in Figure 7.19; this is the basic structure of all the molecules in this series, even for those which belong to a different space group. The Co/Cl (124) and Mn/Cl (125) compounds are orthorhombic, while the Cu/Cl (126) and Mn/Br (125, 127) analogs are monoclinic, with angles β not far from 90°. Chains of edge-sharing octahedra are obtained by means of di-halo bridges; the coordination sphere is trans-$[MCl_4(H_2O)_2]$, and it will be noticed that the O-M-O axis on adjoining octahedra tilt in opposite directions. This has important implications for the magnetic properties, as will be described further below. Furthermore, the chains are linked together, weakly, into planes by another chloride ion, Cl_3, which is hydrogen-bonded to the water molecules in two adjoining chains. The trimethylammonium ions lie between the planes and act to separate them both structurally and magnetically.

It is convenient to begin a discussion of the magnetic properties of this series of compounds by turning first to the monoclinic copper salt,

$[(CH_3)_3NH]CuCl_3 \cdot 2H_2O$. All the significant magnetic interactions are weak and occur below 1 K (126). Both susceptibility (128) and specific heat (129) studies show that long-range order sets in at only about 0.16 K. The principal interactions are ferromagnetic, and in fact the crystal is the only example of a Heisenberg linear chain in which ferromagnetic interactions predominate. The magnetic specific heat is illustrated in Figure 8.28, where curve a is the prediction for a Heisenberg ferromagnetic chain

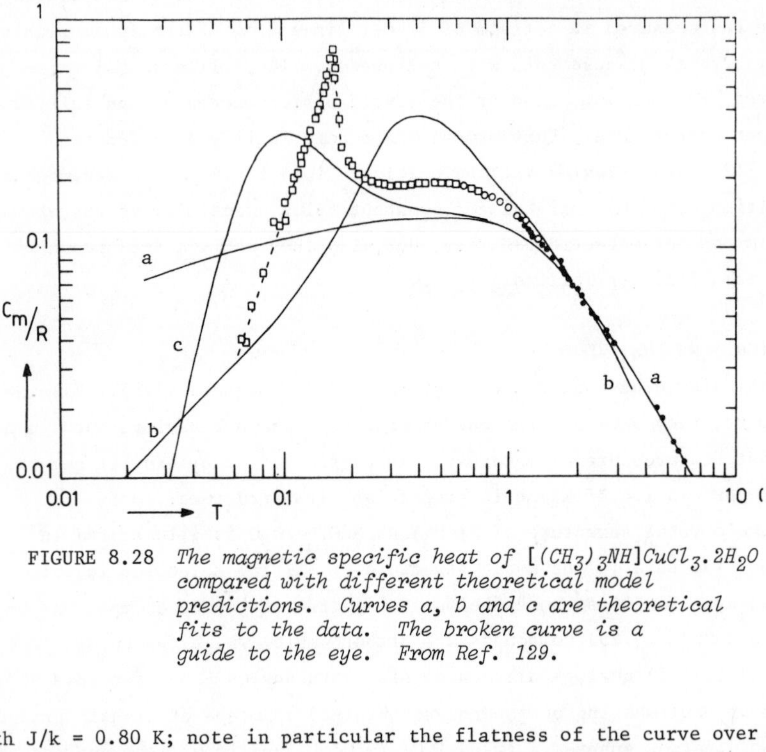

FIGURE 8.28 *The magnetic specific heat of* $[(CH_3)_3NH]CuCl_3 \cdot 2H_2O$ *compared with different theoretical model predictions. Curves a, b and c are theoretical fits to the data. The broken curve is a guide to the eye. From Ref. 129.*

with J/k = 0.80 K; note in particular the flatness of the curve over a large temperature interval, and compare with the theoretical curves in Figure 6.10. Curve b, which does not fit the data adequately, corresponds to the quadratic ferromagnet, while curve c, the best fit, is the prediction for a ferromagnetic chain with slight (Ising) anisotropy, $J_{\parallel}/k$ = 0.85 K, $J_{\perp}/k = 0.9\ J_{\parallel}/k$.

It is interesting to compare the magnetic behavior observed for this compound with that of $CuCl_2 \cdot 2H_2O$ and $CuCl_2 \cdot 2C_5H_5N$. A variety of low-temperature techniques (130) have been used to study the magnetic properties of $CuCl_2 \cdot 2H_2O$. These have revealed a typical antiferromagnet displaying a transition to an ordered state at 4.3 K. The crystal structure (131,132) consists of chains similar to those found in $[(CH_3)_3NH]CuCl_3 \cdot 2H_2O$. These chains are connected by the hydrogen bonds formed between the water molecules of one chain and the chlorine atoms of the neighboring chains. It has been established by NMR (133) and neutron-diffraction (134) studies that the antiferromagnetic transition in $CuCl_2 \cdot 2H_2O$ results in sheets of ferromagnetically ordered spins in the ab planes with antiparallel alignment in adjacent ab planes. The susceptibility measured in the a direction goes through a broad maximum at about 4.5 - 5.5 K (135). Although there is no clear experimental proof of magnetic chain behavior along the c axis other than the presence of short-range order, several calculations have estimated the exchange in this direction. Measurements (136) of the heat capacity suggest substantial short-range order above the Néel temperature since only about 60 % of the magnetic entropy has been acquired at the transition temperature.

In the case of $CuCl_2 \cdot 2C_5H_5N$ (137), square coplanar units aggregate into nearly isolated antiferromagnetic chains by weaker Cu-Cl bonds which complete the distorted octahedron about each copper. The two nonequivalent Cu-Cl bond distances differ more in this structure than those in either $[(CH_3)_3NH]CuCl_3 \cdot 2H_2O$ or $CuCl_2 \cdot 2H_2O$. This compound exhibits (138,139) a broad maximum at 17.5 K in the susceptibility which is characteristic of a magnetic linear chain. Long-range order sets in at 1.13 K. The specific-heat measurements yield a value of $J/k = -13.4$ K.

An interesting feature of these three compounds is the relative magnitude of the exchange within the chemical chains. The values for the exchange parameters are as follows: $J/k < 1$ K for $[(CH_3)_3NH]CuCl_3.2H_2O$, $J/k \approx -7$ K for $CuCl_2.2H_2O$, and $J/k \approx 13$ K for $CuCl_2.2C_5H_5N$, where the values have all been derived from a Heisenberg model. It appears that the compound with the most asymmetry in the $(-CuCl_2-)_n$ unit, i.e., $CuCl_2.2C_5H_5N$, has the largest exchange while the least asymmetric unit, in $[(CH_3)_3NH]CuCl_3.2H_2O$, displays clearly the least amount of exchange along the chemical chain. It is likely that the exchange interactions in $CuCl_2.2H_2O$ causing the transition itself will in some way influence the exchange along the chain. However, it is also perhaps likely that the nearest-neighbor interaction as reflected in the asymmetry will make the largest contribution in determining this exchange.

Now, $[(CH_3)_3NH]CuCl_3.2H_2O$ has a lower transition temperature than $CuCl_2.2H_2O$. If the transition in $CuCl_2.2H_2O$ is assumed to be largely determined by the exchange within the ab plane then the lower transition temperature of the trimethylammonium compound can be hypothesized to be

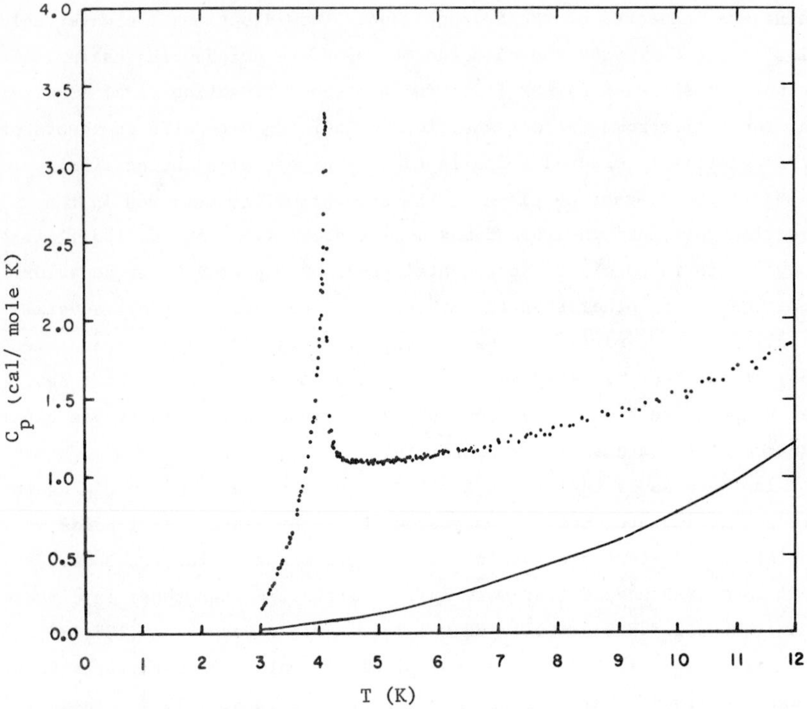

FIGURE 8.29 *Specific heat of* $[(CH_3)_3NH]CoCl_3 \cdot 2H_2O$. *The solid curve is the estimated lattice contribution. From Ref. 124.*

the result of the intervening anionic chloride ions between the chains, which reduce the predominant mode of such exchange. This exchange possibility has been all but eliminated in $CuCl_2 \cdot 2C_5H_5N$ because of the insulating (diluting) effects of the pyridine molecules. Spence (140,141) in recent NMR studies, compared the exchange interactions in $MnCl_2 \cdot 2H_2O$ and $KMnCl_3 \cdot 2H_2O$ and pointed out the extremely delicate balance there is between the geometry and exchange effects. The bond distances within the coordinating octahedron in $MnCl_2 \cdot 2H_2O$ are (Mn-Cl_1) = 2.592 Å, (Mn-Cl_1') = 2.515 Å, and (Mn-O) = 2.150 Å, while in $KMnCl_3 \cdot 2H_2O$ they are (Mn-Cl_1) = 2.594 Å, (Mn-Cl_1') = 2.570 Å, and (Mn-O) = 2.18 Å. From the NMR studies, the spins in $MnCl_2 \cdot 2H_2O$ were found to order antiferromagnetically in chains, while the spins in $KMnCl_3 \cdot 2H_2O$ are ordered ferromagnetically within dimeric units even though both compounds have the same basic octahedral edge-sharing coordination. In any event, very small changes in the coordination geometry are associated with profound changes in the exchange to the point where even the sign of the exchange has been reversed.

It is then reasonable to suggest that small changes in the coordination geometry for this series of copper compounds cause the effects observed in the magnitude of the exchange along the chemical chain.

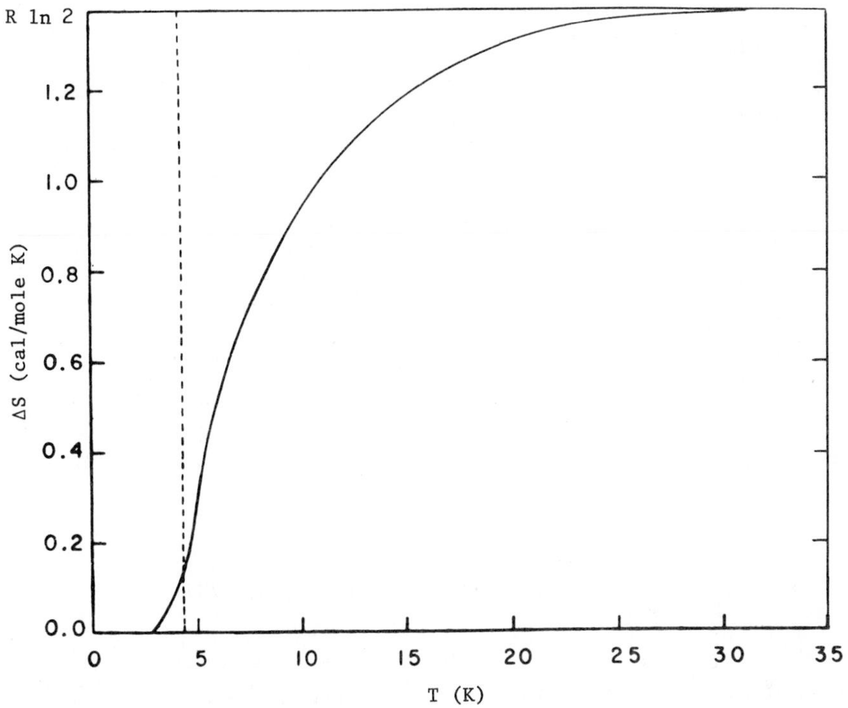

FIGURE 8.30 *Magnetic entropy of* $[(CH_3)_3NH]CoCl_3.2H_2O$ *as a function of temperature. The dashed line indicates the ordering temperature. From Ref. 124.*

The specific heat of $[(CH_3)_3NH]CoCl_3 \cdot 2H_2O$ is illustrated in Figure 8.29, where it will be observed that T_c is about 4.14 K, and that the λ-shaped peak is quite sharp (124). The solid curve is the lattice heat capacity, which was estimated by means of a corresponding states calculation using the measured heat capacity of the copper analog, which is paramagnetic throughout this temperature region. The most significant fact about the measured specific heat is that a mere 8 % of the theoretical maximum entropy for a mole of spin-1/2 Co^{2+} ions, R ln2, has been acquired at the transition temperature. This may be appreciated by looking at

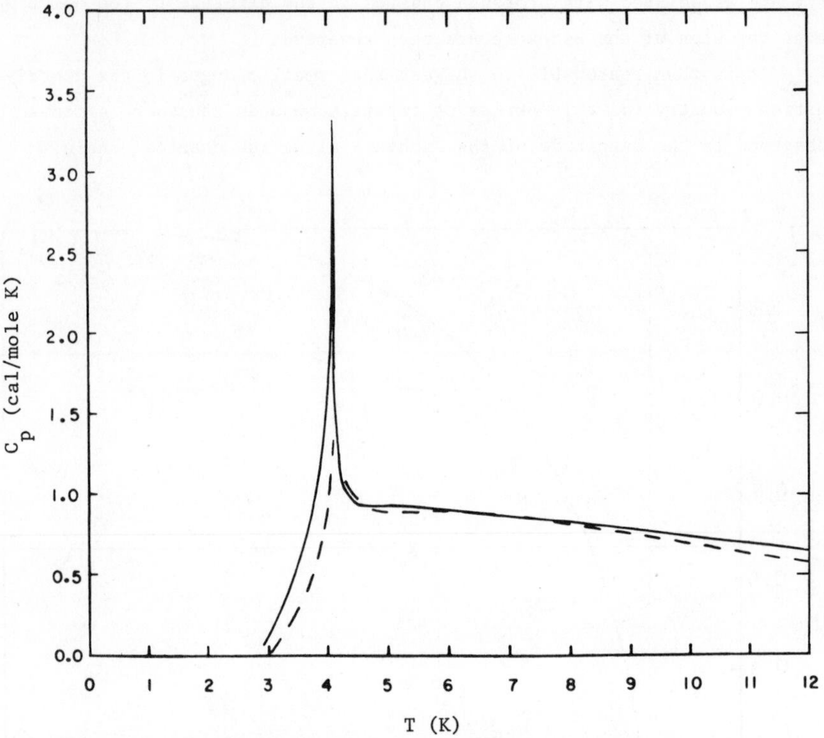

FIGURE 8.31 *Magnetic heat capacity (solid line) of* $[(CH_3)_3NH]CoCl_3 \cdot 2H_2O$. *The dashed line represents the fit to Onsager's equation. From Ref. 124.*

Figure 8.30, a plot of the experimental magnetic entropy as a function of temperature. It is clear that substantial short-range order persists from T_c to well above 15 K in this compound. This short-range order is an indication of the lowered dimensionality of the magnetic spin system, and is consistent with the chain structure. Indeed, the magnetic heat capacity was successfully fitted to Onsager's complete solution (142) for the rectangular Ising lattice, as illustrated in Figure 8.31, with the very anisotropic parameters $|J/k| = 7.7$ K and $|J'/k| = 0.09$ K. Although the sign of the exchange constants is not provided by this analysis, the parameters are consistent with T_c as calculated independently from another equation of Onsager,

$$\sinh(2J/kT_c)\ \sinh(2J'/kT_c) = 1. \tag{8.9}$$

The zero-field susceptibilities (124) provide not only the sign of

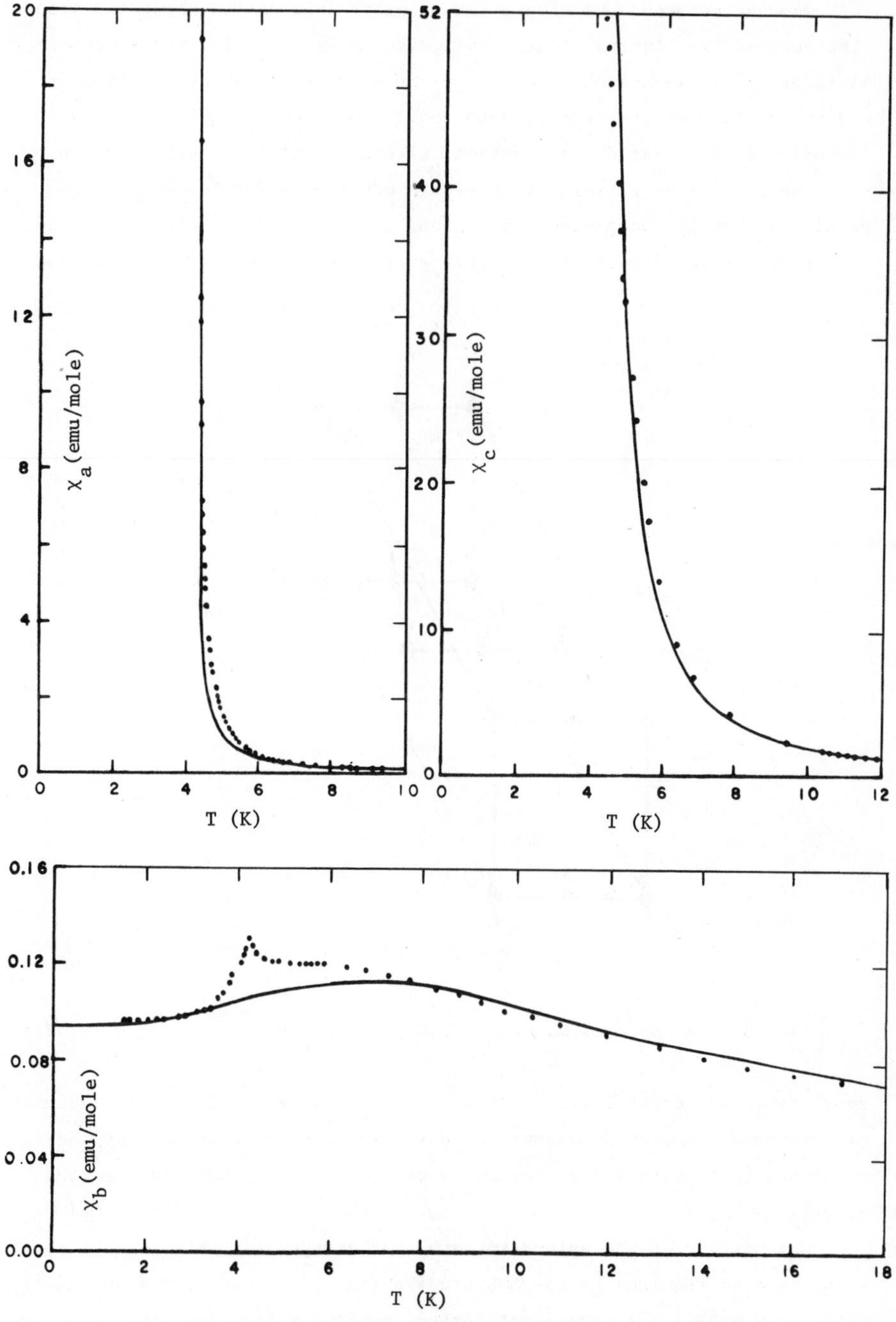

FIGURE 8.32 *Results of the best-fit analyses (solid lines) of the three principal axis magnetic susceptibilities of* $[(CH_3)_3NH]CoCl_3.2H_2O$. *From Ref. 124.*

the exchange constant but also a lot of other information about the spin-structure of this crystal. As illustrated in Figure 8.32, the susceptibilities are quite anisotropic, and χ_b in particular can be fitted over the entire temperature region, both above and below T_c, quite well by Fisher's Ising linear chain equation, with a positive exchange constant.

The a axis susceptibility retains a constant value below T_c, suggesting a weak ferromagnetic moment, and in fact χ_a could only be fitted by requiring a Dzyaloshinsky-Moriya interaction that leads to canting. The

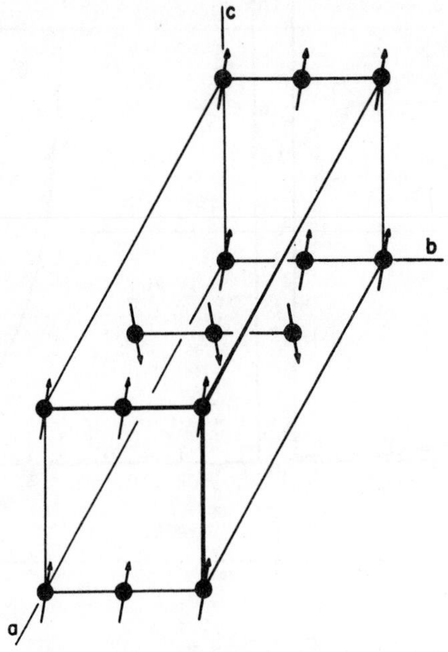

FIGURE 8.33 *Spin-structure of* $[(CH_3)_3NH]CoCl_3 \cdot 2H_2O$. *From Ref. 143.*

symmetry of the crystal allows this, for it will be recalled that a center of inversion symmetry is absent in this lattice because of the way the octahedra in a chain are successively tilted first one way and then the other.

The picture of the spin-structure that emerges for this substance, which is also required by the NMR studies (143), is that ferromagnetically oriented chains are arranged antiferromagnetically, but that the spins are canted in the ac plane so that a net moment persists below T_c. This arrangement is illustrated in Figure 8.33.

The effect of deuterium substitution to form $[(CH_3)_3ND]CoCl_3 \cdot 2D_2O$ on the magnetic properties is quite small (144,145). In fact, T_c is unchanged, within experimental error. This was anticipated because the exchange interaction along the chain should hardly be perturbed by replacing H by D, and the main effect would therefore appear in J'. It happens that because of the weak ferromagnetism of this sample that demagnetization effects which involve both the size and shape of the experimental specimen, limit the experimental accuracy of determination of J'. But, on the other hand, on the application of Eq. (8.9) above, for a constant J/k of the magnitude found here, a 20 % change in the value of the very small J'/k causes only a 4 % change in T_c.

Finally, $[(CH_3)_3NH]CoCl_3.2H_2O$ is also a metamagnet (143), undergoing a phase transition at a mere 64 Oersteds. Interestingly, T_c for the analogous bromide compound is the <u>lower</u> value of 3.86 K (146), and the transition field increases to 120 Oersteds (147). Comparison of the magnetic specific heats of both $[(CH_3)_3NH]CuCl_3 \cdot 2H_2O$ and $[(CH_3)_3NH]CoCl_3 \cdot 2H_2O$ has provided firm evidence (148) for lattice dimensionality crossover effects.

The manganese analogs are similar (125, 127, 149) in both structure and magnetic properties. Broad maxima are observed in the specific heat, which again are evidence of a high degree of short-range order. The susceptibilities indicate that canting also occurs in these molecules, which causes a problem of sorts. As discussed earlier, a canting of spins can be due to either large g-value anisotropy or large zero-field splitting. But, neither of these phenomena is typical of manganese(II), and in fact while $CsCoCl_3 \cdot 2H_2O$ exhibits canting (150), the isomorphic $CsMnCl_3 \cdot 2H_2O$ does not (151). There is no evidence for substantial g value anisotropy with the $[(CH_3)_3NH]MnX_3 \cdot 2H_2O$ (X=Cl,Br) salts, and it is difficult to extract the magnitude of the zero-field splittings. The bromide also goes metamagnetic at the relatively small field of 1200 Oersteds.

REFERENCES

1. C.J. Ballhausen, "Introduction to Ligand Field Theory," McGraw-Hill, New York, 1962.
2. B.R. McGarvey, in "Transition Metal Chemistry," Vol. 3, edited by R.L. Carlin, Marcel Dekker, Inc., New York, 1966, p. 90.
3. F.A. Cotton and G. Wilkinson, "Advanced Inorganic Chemistry," Ed. 3, J. Wiley and Sons, New York, 1972.
4. A. Abragam and B. Bleaney, "Electron Paramagnetic Resonance of Transition Ions," Oxford University Press, Oxford, 1970.

5. J.W. Orton, "Electron Paramagnetic Resonance," Iliffe Books, Ltd., London, 1968.
6. a. F.E. Mabbs and D.J. Machin, "Magnetism and Transition Metal Complexes," Chapman and Hall, London, 1973.
   b. Annual reviews of the literature are provided by the Annual Reports of the Chemical Society, London, and by the Specialist Periodical Reports, also published by the Chemical Society, in the series Electronic Structure and Magnetism of Inorganic Compounds.
7. The early results, as well as the high temperature behavior of $CsTi(SO_4)_2 \cdot 12H_2O$, are reviewed by Mabbs and Machin (6a); the adiabatic demagnetization work is described by D. de Klerk, Handbuch der Physik, XV, Springer, Berlin, p. 100.
8. R.J. Benzie and A.H. Cooke, Proc. Roy. Soc. (London) A209, 269 (1951).
9. B. Bleaney, G.S. Bogle, A.H. Cooke, R.J. Duffus, M.C.M. O'Brien, and K.W.H. Stevens, Proc. Phys. Soc. (London) A68, 57 (1955).
10. J.A. MacKinnon and J.L. Bickerton, Canadian J. Phys. 48, 814 (1970).
11. A. Manoogian, Canadian J. Phys. 48, 2577 (1970).
12. C.T. Prewitt, R.D. Shannon, D.B. Rogers and A.W. Sleight, Inorg. Chem. 8,1985 (1969).
13. R.W. Schwartz and R.L. Carlin, J. Am. Chem. Soc. 92, 6763 (1970).
14. J. Sygusch, Acta Cryst. B30, 662 (1974).
15. N. Rumin, C. Vincent and D. Walsh, Phys. Rev. B7, 1811 (1973).
16. Y.H. Shing, C. Vincent and D. Walsh, Phys. Rev. B9, 340 (1974).
17. Y.H. Shing and D. Walsh, Phys. Rev. Lett. 33, 1067 (1974).
18. Y.H. Shing and D. Walsh, J. Phys. C.: Solid State Phys. 7, L 346 (1974).
19. A. Jesion, Y.H. Shing and D. Walsh, Phys. Rev. Lett. 35, 51 (1975).
20. T.S. Piper and R.L. Carlin, Inorg. Chem. 2, 260 (1963).
21. B.R. McGarvey, J. Chem. Phys. 38, 388 (1963).
22. D.E. Dugdale, J. Phys. C. 1, 1543 (1968).
23. H.M. Gladney and J.D. Swalen, J. Chem. Phys. 42, 1999 (1965).
24. R.L. Carlin, C.J. O'Connor, and S.N. Bhatia, Inorg. Chem. 15, 985 (1976).
25. S. Foner and W. Low, Phys. Rev. 120, 1585 (1960).
26. J. Lambe and C. Kikuchi, Phys. Rev. 118, 71 (1959).
27. Papers presented at the International Conference on Magnetism, Amsterdam, September, 1976, by S.A. Friedberg, J.J. Smit, L.J. de Jongh and R.L. Carlin, see Physica B86-88.
28. A. Syamal, Coord. Chem. Revs. 16, 309 (1975).
29. R.P. Hudson, "Principles and Applications of Magnetic Cooling," North Holland, Amsterdam (1972).
30. O.E. Vilches and J.C. Wheatley, Phys. Rev. 148, 509 (1966).
31. J.R. Sams and T.B. Tsin, Inorg. Chem. 14, 1573 (1975); J.R. Sams and T.B. Tsin, Chem. Phys. 15, 209 (1976).
32. H.A. Goodwin, Coord. Chem. Revs. 18, 293 (1976).
33. R.L. Carlin, Trans. Metal Chem. 1, 1 (1965).
34. C.A. Bates and P.H. Wood, Contemp. Phys. 16, 547 (1975).
35. A. Abragam and M.H.L. Pryce, Proc. Roy. Soc. (London) A206, 173 (1951).
36. K. Kambe, S. Koide, and Y. Usui, Prog. Theor. Phys. 7, 15 (1952).
37. W. Low, Phys. Rev. 109, 256 (1958).
38. B. Bleaney and D.J.E. Ingram, Proc. Roy. Soc. (London) A208, 143 (1951).

39. B.E. Myers, L.G. Polgar, and S.A. Friedberg, Phys. Rev. B6, 3488 (1972).
40. B.E. Myers, L.G. Polgar, and S.A. Friedberg, Phys. Rev. B6, 3488 (1972); Y. Ajiro, S.A. Friedberg and N.S. Vander Ven, Phys. Rev. B12, 39 (1975); S.A. Friedberg, M. Karnezos, and D. Meier, Proc., 14th Conf. on Low Temperature Physics, Otaniemi, Finland, August, 1975, paper L 59.
41. R.L. Carlin, C.J. O'Connor and S.N. Bhatia, J. Amer. Chem. Soc. 98, 3523 (1976).
42. J.N. McElearney, D.B. Losee, S. Merchant, and R.L. Carlin, Phys. Rev. B7, 3314 (1973).
43. F.W. Klaaijsen, Z. Dokoupil and W.J. Huiskamp, Physica 79B, 457 (1975).
44. T Moriya, Phys. Rev. 117, 635 (1960).
45. R.J. Elliott and K.W.H. Stevens, Proc. Roy. Soc. (London) A215, 437 (1952).
46. C.A. Hutchison, Jr., and E. Wong, J. Chem. Phys. 29, 754 (1958).
47. A.H. Cooke, H. Meyer, and W.P. Wolf, Proc. Roy. Soc. (London) A237, 404 (1956).
48. D.A. Langs and C.R. Hare, Chem. Commun. 1967, 890.
49. B.N. Figgis, M. Gerloch, J. Lewis, and R.C. Slade, J. Chem. Soc. 1968, 2028.
50. A.K. Gregson and S. Mitra, J. Chem. Phys. 50, 2021 (1969).
51. J.N. McElearney, D.B. Losee, S. Merchant and R.L. Carlin, J. Chem. Phys. 54, 4585 (1971); J.A. van Santen, Thesis, Leiden, 1978.
52. J.A. van Santen, A.J. van Duyneveldt and R.L. Carlin, Physica 79B, 91 (1975).
53. Y. Numasawa, H. Kitaguchi and T. Watanabe, J. Phys. Soc. Japan 38, 1415 (1975).
54. R. Adams, R. Gaura, R. Raczkowski and G. Kokoszka, Phys. Lett. 49A, 11 (1974).
55. Y. Numasawa and T. Watanabe, J. Phys. Soc. Japan 41, 1903 (1976).
56. Z. Dokoupil, H. den Adel and H.A. Algra, unpublished.
57. F.W. Klaaijsen, H. Suga and Z. Dokoupil, Physica 51, 630 (1971).
58. J.N. McElearney, D.B. Losee, S. Merchant and R.L. Carlin, Phys. Lett. 36A, 129 (1971). Later references are provided by J.E. Rives and V. Benedict, Phys. Rev. B12, 1908 (1975).
59. An early review is included in T. Nagamiya, K. Yosida, and R. Kubo, Adv. Phys. 4, 1 (1955).
60. Gmelin, Handbuch, Vol. 57B, Sect. 2, p. 553 (Fig. 213).
61. W.K. Robinson and S.A. Friedberg, Phys. Rev. 117, 402 (1960).
62. J.N. McElearney, D.B. Losee, S. Merchant and R.L. Carlin, Phys. Rev. B7, 3314 (1973). A portion of Eq. 8 in this paper should be rewritten as

$$\delta_x = [1 - \exp\{-(D+E)/kT\}]/Z_o(D + E)$$

and

$$\delta_y = [1 - \exp\{-(D-E)/kT\}]/Z_o(D - E)$$

63. L.G. Polgar, A. Herweijer and W.J.M. de Jonge, Phys. Rev. B5, 1957 (1972).

64. H. Bizette, Compt. rend. 243, 1295 (1956).
65. S.N. Bhatia and R.L. Carlin, Physica 86-88B, 903 (1977); S.N. Bhatia, R.L. Carlin, and A.P. Filho, Physica B, in press.
66. K. Kopinga and W.J.M. de Jonge, Phys. Lett. 43A, 415 (1973).
67. L.G. Van Uitert, H.J. Williams, R.C. Sherwood and J.J. Rubin, J. Appl. Phys. 36, 1029 (1965).
68. T. Haseda, H. Kobayashi and M. Date, J. Phys. Soc. Japan 14, 1724 (1959).
69. J. Mizuno, J. Phys. Soc. Japan 16, 1574 (1961).
70. A.I. Hamburger and S.A. Friedberg, Physica 69, 67 (1973).
71. R. Kleinberg, J. Chem. Phys. 50, 4690 (1969).
72. R. Kleinberg, J. Appl. Phys. 38, 1453 (1967).
73. C.C. Becerra and A.P. Filho, Phys. Lett. 44A, 13 (1973); A.P. Filho, C.C. Becerra and N.F. Oliveira, Jr., Phys. Lett. 50A, 51 (1974); N.F. Oliveira, Jr., A.P. Filho, and S.R. Salinas, Phys. Lett. 55A, 293 (1975).
74. A.L.M. Bongaarts, B. van Laar, A.C. Botterman and W.J.M. de Jonge, Phys. Lett. A41, 411 (1972); C.H.W. Swüste, A.C. Botterman, J. Millenaar, and W.J.M. de Jonge, J. Chem. Phys., 66, 5021 (1977).
75. T. de Neef and W.J.M. de Jonge, Phys. Rev. B10, 1059 (1974).
76. L. Berger and S.A. Friedberg, Phys. Rev. 136, A158 (1964).
77. V.A. Schmidt and S.A. Friedberg, Phys. Rev. B1, 2250 (1970).
78. L.G. Polgar and S.A. Friedberg, Phys. Rev. B4, 3110 (1971).
79. A. Herweijer and S.A. Friedberg, Phys. Rev. B4, 4009 (1971).
80. R.L. Martin and A.H. White, Trans. Met. Chem. 4, 113 (1968).
81. M.E. Switzer, R. Wang, M.F. Rettig, and A.H. Maki, J. Am. Chem. Soc. 96, 7669 (1974).
82. L.H. Pignolet, G.S. Patterson, J.F. Weiher, and R.H. Holm, Inorg. Chem. 13, 1263 (1974).
83. P.B. Merrithew and P.G. Rasmussen, Inorg. Chem. 11, 325 (1972).
84. G. Harris, Theor. Chim. Acta 5, 379 (1966); 10, 119 (1968).
85. G.R. Hall and D.N. Hendrickson, Inorg. Chem. 15, 607 (1976).
86. J.G. Leipoldt and P. Coppens, Inorg. Chem. 12, 2269 (1973).
87. B.F. Hoskins and C.D. Pannan, Inorg. Nucl. Chem. Lett. 11, 409 (1975).
88. R.L. Martin and A.H. White, Inorg. Chem. 6, 712 (1967).
89. B.F. Hoskins and A.H. White, J. Chem. Soc. A1970, 1668.
90. P.C. Healey, A.H. White, and B.F. Hoskins, J. Chem. Soc. D 1972, 1369.
91. G.E. Chapps, S.W. McCann, H.H. Wickman, and R.C. Sherwood, J. Chem. Phys. 60, 990 (1974).
92. G.C. Brackett, P.L. Richards, and W.S. Caughey, J. Chem. Phys. 54, 4383 (1971).
93. H.H. Wickman and F.R. Merritt, Chem. Phys. Lett. 1, 117 (1967).
94. P. Ganguli, V.R. Marathe, and S. Mitra, Inorg. Chem. 14, 970 (1975).
95. G.C. DeFotis, F. Palacio, and R.L. Carlin, to be published.
96. H.H. Wickman, A.M. Trozzolo, H.J. Williams, G.W. Hull and F.R. Merritt, Phys. Rev. 155, 563 (1967); 163, 526 (1967).
97. H.H. Wickman and A.M. Trozzolo, Inorg. Chem. 7, 63 (1968).
98. H.H. Wickman and C.F. Wagner, J. Chem. Phys. 51, 435 (1969).
99. H.H. Wickman, J. Chem. Phys. 56, 976 (1972).
100. J.M. Grow and H.H. Wickman, AIP Conf. Proc. #24, 215 (1974); A. Kostikas, D. Petrides, A. Simopoulos, and M. Pasternak, Solid State Comm. 13, 1661 (1973); J.M. Grow, G.L. Robbins, and H.H. Wickman, Chem. Phys. Lett. 43, 77 (1976).

101. P. Burlet, P. Burlet and E.F. Bertaut, Solid State Comm. 14, 665 (1974).
102. E.F. Bertaut, T.Q. Duc, P. Burlet, P. Burlet, M. Thomas, and J.M. Moreau, Acta Crystallogr. B30, 2234 (1974).
103. R.B. Flippen and S.A. Friedberg, Phys. Rev. 121, 1591 (1961).
104. J.H. Schelleng, C.A. Raquet and S.A. Friedberg, Phys. Rev. 176, 708 (1968).
105. R.D. Spence, J. Chem. Phys. 62, 3659 (1975).
106. V.A. Schmidt and S.A. Friedberg, Phys. Rev. 188, 809 (1969).
107. A.D. van Ingen Schenau, G.C. Verschoor, and C. Romers, Acta Cryst. B30, 1686 (1974).
108. T.J. Bergendahl and J.S. Wood, Inorg. Chem. 14, 338 (1975).
109. H.A. Algra, L.J. de Jongh, W.J. Huiskamp and R.L. Carlin, Physica 83B, 71 (1976).
110. R.L. Carlin, C.J. O'Connor and S.N. Bhatia, J. Am. Chem. Soc. 98, 685 (1976).
111. C.J. O'Connor and R.L. Carlin, Inorg. Chem. 14, 291 (1975).
112. H.A. Algra, L.J. de Jongh, and R.L. Carlin, to be published.
113. H.A. Algra, L.J. de Jongh, and R.L. Carlin, to be published.
114. H.A. Algra, L.J. de Jongh, and R.L. Carlin, to be published.
115. M. Tachiki and T. Yamada, J. Phys. Soc. Japan 28, 1413 (1970); M. Tachiki, T. Yamada, and S. Maekawa, J. Phys. Soc. Japan 29, 656 (1970); T. Tsuneto and T. Murao, Physica 51, 1861 (1971).
116. K.M. Diederix, H.A. Algra, J.P. Groen, T.O. Klaassen, N.J. Poulis, and R.L. Carlin, Phys. Lett. 60A, 247 (1977).
117. H.A. Algra, J. Bartolome, K.M. Diederix, L.J. de Jongh, and R.L. Carlin, Physica 85B, 323 (1977).
118. H.A. Algra, F.J.A.M. Greidanus, L.J. de Jongh, W.J. Huiskamp, and J. Reedijk, Physica 83B, 85 (1976).
119. R.L. Carlin, H.A. Algra, C.J. O'Connor and S.N. Bhatia, unpublished.
120. H.T. Witteveen, W.L.C. Rutten, and J. Reedijk, J. Inorg. Nucl. Chem. 37, 913 (1975).
121. F.W. Klaaijsen, Thesis, Leiden, 1974; F.W. Klaaijsen, H.W.J. Blöte, and Z. Dokoupil, Solid State Comm. 14, 607 (1974); F.W. Klaaijsen, Z. Dokoupil, and W.J. Huiskamp, Physica 79B, 547 (1975). Similar studies on the manganese analogs are published by F.W. Klaaijsen, H.W.J. Blöte, and Z. Dokoupil, Physica 81B, 1 (1976).
122. H.W.J. Blöte, Physica 79B, 427 (1975).
123. J.N. McElearney, G.E. Shankle, D.B. Losee, S. Merchant and R.L. Carlin, ACS Symposium Series, Number 5, "Extended Interactions Between Metal Ions in Transition Metal Complexes," Ed. L.V. Interrante, American Chemical Society, Washington, D.C., 1974, p. 194.
124. D.B. Losee, J.N. McElearney, G.E. Shankle, R.L. Carlin, P.J. Cresswell and W.T. Robinson, Phys. Rev. B8, 2185 (1973).
125. R.E. Caputo, R.D. Willett, and J.A. Muir, Acta Crystallogr. B32, 2639 (1976).
126. D.B. Losee, J.N. McElearney, A. Siegel, R.L. Carlin, A.A. Khan, J.P. Roux and W.J. James, Phys. Rev. B6, 4342 (1972).
127. S. Merchant, J.N. McElearney, G.E. Shankle, and R.L. Carlin, Physica 78, 308 (1974).
128. C.R. Stirrat, S. Dudzinski, A.H. Owens and J.A. Cowen, Phys. Rev. B9, 2183 (1974).
129. H.A. Algra, L.J. de Jongh, H.W.J. Blöte, W.J. Huiskamp and R.L. Carlin, Physica 78, 314 (1974).
130. C.J. Gorter, Rev. Mod. Phys. 25, 332 (1953).

131. D. Harker, Z. Krist. 93, 136 (1936).
132. A. Engberg, Acta Chem. Scand. 24, 3510 (1970).
133. N.J. Poulis and G.E.G. Hardeman, Physica 18, 201 (1952); 18, 315 (1952); 19, 391 (1953).
134. G. Shirane, B.C. Frazer and S.A. Friedberg, Phys. Lett. 17, 95 (1965).
135. L.C. van der Marel, J. van den Broek, J.D. Wasscher and C.J. Gorter, Physica 21, 685 (1955).
136. S.A. Friedberg, Physica 18, 714 (1952).
137. J.D. Dunitz, Acta Cryst. 10, 307 (1957).
138. K. Takeda, S. Matsukawa and T. Haseda, J. Phys. Soc. Japan 30, 1330 (1971).
139. W. Duffy, Jr., J.E. Venneman, D.L. Strandburg and P.M. Richards, Phys. Rev. B9, 2220 (1974).
140. R.D. Spence and K.V.S. Rama Rao, J. Chem. Phys. 52, 2740 (1970).
141. R.D. Spence, W.J.M. de Jonge and K.V.S. Rama Rao, J. Chem. Phys. 54, 3438 (1971).
142. L. Onsager, Phys. Rev. 65, 117 (1944).
143. R.D. Spence and A.C. Botterman, Phys. Rev. B9, 2993 (1974).
144. R.L. Carlin, C.J. O'Connor and S.N. Bhatia, Phys. Lett. 50A, 433 (1975).
145. S.N. Bhatia, C.J. O'Connor and R.L. Carlin, Inorg. Chem. 15, 2900 (1976).
146. J.N. McElearney, unpublished.
147. R.D. Spence, unpublished.
148. H.A. Algra, L.J. de Jongh, W.J. Huiskamp, and R.L. Carlin, Physica, to be published.
149. P.R. Newman, J.A. Cowen and R.D. Spence, AIP Conference Proceedings No. 18 (1974), p. 391.
150. A. Herweijer, W.J.M. de Jonge, A.C. Botterman, A.L.M. Bongaarts and J.A. Cowen, Phys. Rev. B5, 4618 (1972).
151. H. Kobayashi, I. Tsujikawa and S.A. Friedberg, J. Low Temp. Phys. 10, 621 (1973).

APPENDIX

A. PHYSICAL CONSTANTS *

Molar gas constant	$R = 8.3144\ \mathrm{J\ mol^{-1}\ K^{-1}}$
Avogadro constant	$N = 6.0220 \times 10^{23}\ \mathrm{mol^{-1}}$
Boltzmann constant	$k = 1.3807 \times 10^{-16}\ \mathrm{erg\ K^{-1}}$
Bohr magneton	$\mu_B = 9.274 \times 10^{-21}\ \mathrm{erg\ G^{-1}}$

Easy to remember

$$N\mu_B^2/k \approx 0.375\ \mathrm{emu\ K\ mol^{-1}}$$

and for the translation of energy 'units'

$$1\ \mathrm{cm^{-1}} \approx 30\ \mathrm{Gc/s} \approx 1.44\ \mathrm{K} \approx 1.24 \times 10^{-4}\ \mathrm{eV} \approx 1.99 \times 10^{-16}\ \mathrm{erg}.$$

B. HYPERBOLIC FUNCTIONS

The hyperbolic functions occur repeatedly in the theory of magnetism. Though they are described in most elementary calculus texts, some properties are summarized here.

The two basic hyperbolic functions are defined in terms of exponentials as follows

$$\sinh x = 1/2\ (\ \exp(x) - \exp(-x)\ )$$

* From Physics Today, Sept., 1974.

$$\cosh x = 1/2\ (\ \exp(x) + \exp(-x)\ )$$

and, by analogy to the common trigonometric functions, there are four more hyperbolic functions defined in terms of sinh x and cosh x, as follows

$$\tanh x = \sinh x\ /\ \cosh x$$
$$\operatorname{sech} x = 1/\cosh x$$
$$\coth x = \cosh x\ /\ \sinh x$$
$$\operatorname{csch} x = 1/\sinh x$$

These functions are sketched in Figure A.1. It is evident that, unlike the trigonometric functions, none of the hyperbolic functions is periodic. Sinh x, tanh x, coth x and csch x are odd functions, while cosh x and

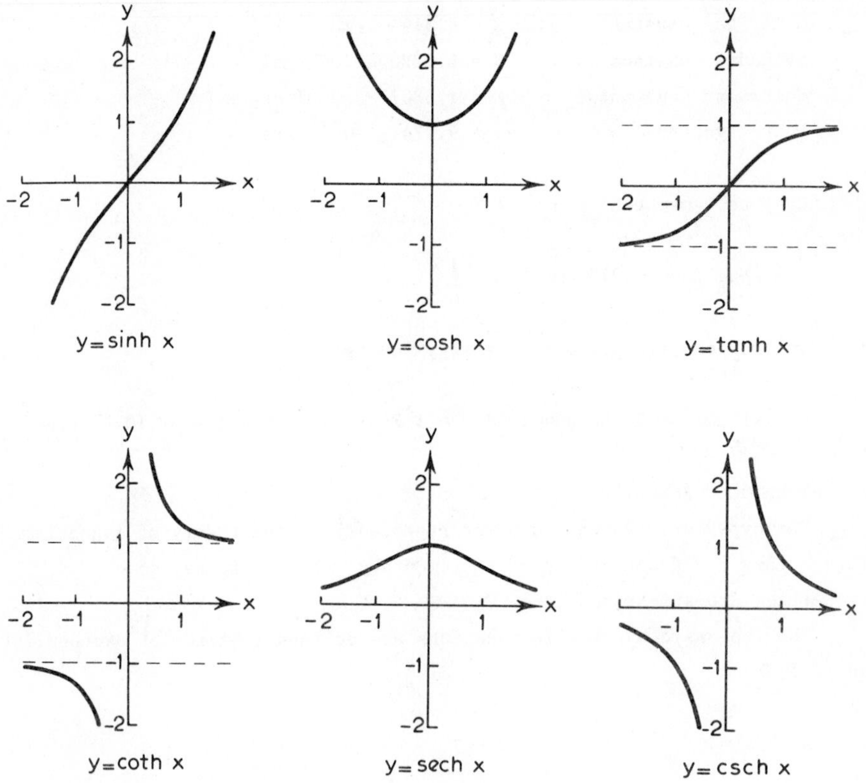

FIGURE A.1 *The hyperbolic functions.*

sech x are even.

Lastly, the derivatives of the hyperbolic functions may be shown to be,

$$d \sinh x = \cosh x \, dx$$

$$d \cosh x = \sinh x \, dx$$

$$d \tanh x = \operatorname{sech}^2 x \, dx$$

$$d \coth x = - \operatorname{csch}^2 x \, dx$$

$$d \operatorname{sech} x = - \operatorname{sech} x \tanh x \, dx$$

$$d \operatorname{csch} x = - \operatorname{csch} x \coth x \, dx$$

SUBJECT INDEX

FORMULA INDEX